Eric Robinson

Department of History,
University of Manchester.

ENGINEERING AT CAMBRIDGE UNIVERSITY 1783–1965

ENGINEERING AT CAMBRIDGE UNIVERSITY 1783–1965

T. J. N. HILKEN

Secretary of the University Department of Engineering

Fellow and Vice-President of University College

CAMBRIDGE AT THE UNIVERSITY PRESS 1967

Published by the Syndics of the Cambridge University Press
Bentley House, 200 Euston Road, London, N.W. 1
American Branch: 32 East 57th Street, New York, N.Y. 10022

Library of Congress Catalogue Card Number: 67-10043

Printed in Great Britain
at the University Printing House, Cambridge
(Brooke Crutchley, University Printer)

CONTENTS

AUTHOR'S PREFACE

I started the research which led to the writing of this book in 1957 at the suggestion of Professor Sir John Baker, for whose help and encouragement I am most grateful. The book tells how engineering entered the university towards the end of the eighteenth century as the hobby of a professor whose statutory duty was to find a cure for the gout, and how it gradually came to be accepted first as part of a liberal education for men reading for the Ordinary Degree and later as a fully fledged subject for university teaching and research. It explains how engineering fits into the general pattern of a Cambridge education, and gives some indication of the special contribution which this university makes to the development of higher technology in Britain. I hope that it will prove interesting not only to graduates of this department, but to educational authorities and others who are concerned with engineering as a university subject, to directors of research in industry and careers masters in schools. It takes the story up to the end of 1965. Anyone wishing for further or more up-to-date information about the current courses or facilities for research should write to The Secretary, University Department of Engineering, Trumpington Street, Cambridge.

T. J. N. H.

Cambridge
September 1966

ACKNOWLEDGEMENTS

During my early researches into the university background to this department I received a great deal of help from Dr J. P. C. Roach of Corpus Christi College, editor of the Cambridgeshire volume of the *Victoria County History* and author of the section within it which deals with the history of the university. I am also most grateful to Mr John Saltmarsh of King's College for the interest he has shown in my work and for reading and criticizing some of the earlier manuscripts.

Dr D. T. Whiteside kindly made available to me information based on his research work on Newton's manuscripts, and I used this as the basis of the opening section. For information about the Hopkinson family and the life of the Department under Professor Bertram Hopkinson I am indebted to Mrs Bertram Hopkinson and her daughter, Dr Phyllis Roughton. Mrs D'O. R. Schofield, daughter of Sir Charles Inglis, kindly allowed me to make use of some of her father's private papers. Professor Sir Melvill Jones helped me with the section describing the establishment of the Professorship of Aeronautical Engineering. Sir Richard Southwell and Mr A. L. Bird read the Ewing and Inglis chapters and made valuable suggestions from first-hand knowledge of the events which I have described. I have also received much help from members of the departmental teaching staff, particularly the professors, and secretaries of the Faculty Board, who advised me on the sections of the last chapter dealing with research and teaching. I am indebted to Mr K. A. Knell, the departmental librarian, for his help and advice in preparing the bibliography and index and to Mr E. R. Mudd for drawing the plans of the buildings and the graph.

Finally I should like to express my gratitude to my wife for listening so patiently while various drafts were read aloud to her and for her many helpful suggestions for improving them, and to my son Richard for his ruthless elimination of clichés from the sections which I passed to him for comment.

T.J.N.H.

LIST OF PLANS

1 THE UNIVERSITY BACKGROUND TO THE STUDY OF ENGINEERING

THE NEWTONIAN TRADITION

The mechanical sciences have been studied and taught in the University of Cambridge without a break since Isaac Milner gave his first course of lectures on Experimental Philosophy in 1783. He introduced them to illustrate the practical applications of Newtonian mathematics, which dominated undergraduate teaching from the beginning of the eighteenth century to the middle of the nineteenth.

In the Middle Ages the main task of universities was the education of the clergy, who provided the administrators and civil servants of Western Europe as well as the ministers of the Church. After the Reformation civil administration in England fell largely into the hands of laymen, and although most of the undergraduates at Oxford and Cambridge continued for another three hundred years to take Holy Orders the education of laymen assumed increasing importance. Well-born young men came to the Universities to enjoy the 'delight and ornament' of the new learning, and the best of them afterwards formed an intellectual class which was able and willing to support works of genius in both the arts and the sciences.

In the 1620s the number of undergraduates at Cambridge was higher than it had ever been; over thirteen hundred were in residence in the colleges, a total which was not reached again for two hundred years. The Civil War found the university mainly on the side of the king, although a powerful Puritan minority favoured the Parliament. In 1644 and again after the king's execution a drastic purge of the Royalists took place, but teaching continued throughout the Protectorate and numbers were maintained at a surprisingly high level. The classics, and in particular the works of Aristotle, provided the main subjects of study, though there were already critics who bewailed the slavish dependence of the universities on the teachings of antiquity.

When the Royalists returned to power after the Restoration Oxford adopted the Anglicanism of Laud and the politics of the Tories, and theology, classical philosophy and political history became the main subjects of study. Cambridge, though loyal to the Crown, was less ready to bow to the dictates of authority. For many years one of the most influential bodies in the university was a group which became known as the Cambridge Platonists, who condemned all religious dogmatism and exalted the freedom of the spirit. Their philosophy was metaphysical rather than mechanical, but they were not entirely unsympathetic to the growing interest in scientific research. Mathematics, which had earlier been an important subject of study at Oxford, entered the Cambridge curriculum soon after the Civil War, and in 1663 Henry Lucas, who had represented the university in Parliament, founded a Professorship of Mathematics. Unlike the holders of most of the earlier chairs the Lucasian Professor was not obliged to take Holy Orders, and to ensure that he paid proper attention to his duties he was forbidden to accept a cure of souls and was required to reside in the university.

At about the same time the works of the French philosopher René Descartes began to receive serious attention in Cambridge. Descartes was the first thinker of modern times to attempt the formulation of a system, independent of the scholastic tradition or the dogmas of theology, which would explain the workings of the whole material universe as a simple mechanism governed by the laws of mathematics. After a preliminary essay entitled *Discourse on Method* he developed his ideas in three great treatises—the *Dioptric*, the *Meteors* and the *Geometry*. From 1665 to 1685 his theories formed a major subject of study, discussion and argument in the university.

The first Lucasian Professor was Isaac Barrow, a Fellow of the Royal Society and a former Regius Professor of Greek who at the time of his election was holding the chair of geometry at Gresham College, London. Although Barrow had been brought up in the old traditions he was receptive of new ideas and was widely regarded as one of the leading mathematicians of the day. He held the chair with distinction for six years and then resigned it in favour of a junior member of his college called Newton.

Isaac Newton, the posthumous son of a Lincolnshire farmer, was born at Woolsthorpe in the parish of Colsterworth on Christmas Day 1642. His mother married again two years later and moved to her husband's vicarage in the neighbouring village of North Witham, and Isaac was placed in the charge of his maternal grandmother. After attending two small local schools he was sent to the grammar school at Grantham, where he lodged in an attic over an apothecary's shop in the High Street kept by a man named Clark. He was temporarily recalled to the farm at Woolsthorpe about 1656, after his mother had been widowed for the second time, but he soon returned to Grantham to resume his studies. Nothing definite is known about the curriculum there, but it certainly included a sound grounding in Latin and Greek grammar, literature and history and the standard rules of arithmetic. Algebra, trigonometry and geometry were then all adult studies, not taught to schoolboys, but he may have obtained some slight acquaintance with them when browsing through a collection of books on scientific subjects stored in his attic by Clark.

In June 1661, when he was nineteen and a half years old, Newton was admitted to Trinity College, Cambridge, as a sizar, a title applied to undergraduates who paid lower fees than the normal and performed certain menial duties in return for their board and instruction. From this time our knowledge of Newton's intellectual development is based on the evidence of a mass of his personal papers, most of which have been published. On his arrival in Cambridge as a freshman he was diffident and socially inexperienced, and has been described as 'a wholly unsophisticated provincial puritan'. He must, however, have impressed the college authorities with his ability as at the end of his third year he was awarded a scholarship.

The undergraduate syllabus in the seventeenth century was still based largely on the medieval 'trivium', a three-year course in logic, ethics and rhetoric. It was severely scholastic in structure, and was concerned mainly with the writings of Aristotle and his commentators. Teaching was by lecture and disputation, the latter being conducted in Latin, with arguments and counter arguments laid out point by point according to the formal laws of logic. This rigid discipline doubtless provided an excellent mental training, but it left little scope for the imagination. It was not until his fourth year as an undergraduate that Newton was able to take full advantage of the opportunities offered for reading outside the scholastic curriculum, and it was then that his attention was directed, probably for the first time, to the serious study of mathematics and physics. The effect was electric, and the summer of 1664 saw the first stirrings of his stupendous creative genius.

It was in this year that the Great Plague reached Cambridge. The university was closed from September to the following March and again from August until February 1666, but Newton remained in residence until May 1665, having taken his B.A. degree in January. He returned in March 1666 but retreated again in June and stayed away until April 1667. His whereabouts during this period are not certainly known, but he probably spent most of his time with his mother, who was still living at North Witham. The family farm was let to tenants, and it is unlikely that Newton lived there, though in his old age he was fond of telling the story of the falling apple in the Woolsthorpe orchard, which he attributed to this period of enforced rustication.

Wherever he was he continued his studies with unabated energy and his notebooks show the extent of his reading and the depth of his researches. The works of Descartes and Kepler on optics and of Gassendi and Streete on astronomy made him aware of the need for a greater knowledge of advanced mathematics and he turned his attention first to Euclid and then to Descartes' *Geometry*. Almost immediately he leapt far ahead of his teachers. His first calculus researches date from the summer of 1664 and by the late autumn he was working on the binomial theorem. When he returned to Cambridge in the spring of 1667 he had mastered all the existing mathematical techniques and perfected his new system of fluxions, or the calculus. He now possessed the tool which enabled him in the years which followed to complete the task which had defeated Descartes and evolve a mathematical theory of general application to the movements of the material universe.

Meanwhile his position in the university was consolidated. He was elected to a minor fellowship by his College in October 1667 and to a major one in March 1668, and in the following July he was admitted to the M.A. degree. In 1669, as already stated, he succeeded Barrow as Professor of Mathematics. In the early years of his professorship he gave courses of lectures on optics and some, if not all, of his lectures were given in Great St Mary's, the university church, which since it was rebuilt at the beginning of the sixteenth century had often been used for occasions of this kind. Lecturing, however, was a relatively minor activity. Most of his time was devoted to intense application to the problems of natural science, and as his studies and experiments proceeded he gradually evolved the theory of gravitation and applied it to the motions of the moon and the planets. In 1672 he was elected to a fellowship in the Royal Society and this brought him into contact or correspondence with most of the great scientific thinkers of the age.

The Royal Society had its origins in a group of amateur scientists who met in London from 1645 onwards to study 'Experimental Philosophy'. In 1660 a formal constitution was approved by Charles II, and two years later the first charter of incorporation passed the Great Seal. Christopher Wren and Robert Boyle were among the founder members and Robert Hook was the first Curator of Experiments and later Secretary. The Society kept up an active correspondence with similar bodies on the continent, including the Académie des Sciences in France, and selections from this correspondence formed the basis of the *Philosophical Transactions* which first appeared under that title in 1665.

In the summer of 1684 Edmund Halley, the future Astronomer Royal, paid Newton a visit which had far-reaching consequences. Halley had already published a paper on the planetary orbits, and the immediate purpose of his visit was to inquire whether Newton could prove that elliptical planetary motion was a theoretical consequence of the inverse square law. Their conversation, however, ranged over much wider fields and Halley was surprised to learn how much progress his host had made in the researches on astronomical theory which had occupied him during the preceding four years. He persuaded him to expand his first results into a short tract, the *De Motu*, which formed the basis of his lectures during the Michaelmas Term and was registered at the Royal Society in the following February. This proved to be the first step towards the writing of the far greater work—probably the greatest scientific treatise of all time—the *Philosophiae Naturalis Principia Mathematica*, the first volume of which was ready for the press in the summer of 1686. Three months later he completed the second book and continued at once with the third, in which the mathematical theories described in the first were applied to the chief phenomena of the solar system. This involved a vast number of new calculations, but he completed the manuscript in 1686, and although the printing proved to be a slow and difficult business, the great work was published in the summer of 1687. Being written in Latin it was available without translation to learned men of all nationalities.

In the autumn of 1692 Newton had what would now be called a nervous breakdown which lasted for about two months. He suffered terribly from insomnia and general nervous irritability, but in the end he made a good recovery and resumed his mathematical research, much of which is still unpublished. In 1695 he was appointed Warden of the Mint and given the task of reminting the national coinage. He handed over his university duties to a deputy, but in 1701 he resigned both from the Lucasian chair and from his fellowship at Trinity. In the following year he was elected President of the Royal Society, a post which he held till his death. In 1709 he allowed Roger Cotes, the first Plumian Professor of Astronomy and Experimental Philosophy, to prepare the second edition of the *Principia*, which was published in 1713; a third edition under the direction of Henry Pemberton was issued in 1726. Newton died on 20 March 1727 and was buried with great pomp in Westminster Abbey. Pope wrote a couplet for his epitaph

> Nature and Nature's laws lay hid in night:
> God said, Let Newton be! and all was light.

The rise of the Newtonian school in Cambridge dates from the publication of the *Principia*. At first it was Newton's theory of the universe which excited most attention in the university, but because mathematics provided the key to that theory it soon became the predominant subject of study in the colleges. One of the first men to appreciate the importance of Newton's work was Richard Laughton, who in 1694, as senior tutor of Clare Hall, persuaded one of his pupils to defend in the Schools a proposition from the *Principia* instead of the orthodox quotation from Aristotle. Later this same pupil, Samuel Clark, wrote the first known textbook on the Newtonian philosophy. The acceptance of Newton's theories as the basis for undergraduate teaching was, however, largely due to Richard Bentley, the dictatorial

Master of Trinity about whose war with his Fellows so much has been written. From his election to the Mastership in 1699 until his death in 1742 he was the outstanding personality in the university. He was a tyrant who brooked no opposition and had no respect for laws or customs, but he was a great reformer and an all-round scholar of exceptional ability. His successor at Trinity was the Plumian Professor Robert Smith, an authority on optics, hydrostatics and harmonics. He endowed two prizes for mathematics and natural philosophy which, with the position of Senior Wrangler, were for over a century the supreme objects of the ambition of the cleverest undergraduates in their final year.

The medieval system of examination by disputation in Latin, which will be described in the next section, was quite unsuited to test the knowledge of students in the new curriculum, and about 1730 a written examination in the Senate House was instituted to establish the order of merit among the students for the B.A. degree. After 1763 the Senate House Examination was accepted as the final test by which a man's place in the list was determined, but the Latin disputations continued to be held until 1839 and were regarded by many as a particularly valuable feature of the Cambridge system of education.

In spite of the attractions offered to men of intelligence by the vast fields opened up by Newton's discoveries, Georgian England took little interest in university education. Matriculations had been falling steadily since the days of Charles II and they reached their nadir in the 1760s, when the total undergraduate population dropped below four hundred, of whom about a third were members of Trinity and St John's. University teaching was practically non-existent, as the professors, finding that their lectures attracted little attention, ceased for the most part to lecture at all. In the colleges the teaching, which in the early days had been shared by most of the fellows, was now confined entirely to one or two tutors. The Master and Bursar were able to occupy themselves with college administration, but the other resident fellows, having no specific duties assigned to them, often degenerated into a collection of uncouth and unpleasant eccentrics. In most of the colleges ordination was compulsory for fellows within a fixed period after election, but in others some fellowships were invariably held by laymen. The latter included Trinity, where Newton held a lay fellowship. One of the statutes under the Elizabethan code which was rigidly enforced in all the colleges on clerical and lay fellows alike was that which required the surrender of the fellowship on marriage. Few of the dons had private means, and those who had been ordained found themselves trapped in the college for life unless a vacancy in a college living enabled them to marry and make their escape to a town or country parish. Some of them took mistresses and raised families in the neighbouring villages outside the university precincts, and there were well-established colonies of this kind in Chesterton and Grantchester. In many ways the record of the mid-eighteenth century forms the least edifying chapter in the long history of the university.

UNDERGRADUATE TEACHING IN THE EIGHTEENTH CENTURY

During the latter half of the eighteenth century the university gradually shook off the sloth and apathy which characterized it during the earlier years. From about 1770 the annual admissions crept slowly upwards, and dons and undergraduates alike began to show a more genuine desire for knowledge. Those colleges which had good tutors attracted good students, and as the Senate House Examination became better organized competition for the Senior Wranglerships and Smith's prizes increased and the general standard of scholarship improved. John Jebb of Peterhouse made great efforts in the seventies to widen the curriculum and spur the undergraduates to greater efforts by instituting annual examinations, and although he was defeated by the forces of conservatism his criticisms of the prevailing educational system were not entirely without effect. The spirit of change and reform was abroad in western Europe, and the outbreak of the French Revolution was welcomed by liberal minded people in Cambridge as elsewhere. Inevitably, however, the excesses of the Terror brought about a reaction, and the so-called 'Jacobins' remained suspect in the university throughout the Napoleonic Wars.

As in the Middle Ages the undergraduates were divided into four classes—noblemen, fellow-commoners, pensioners and sizars. The first two paid higher fees than the others, dined at the high tables, and on ceremonial occasions wore splendid gowns of purple, white, green or rose embroidered with gold and silver. Noblemen were admitted to the M.A. degree without examinations after six terms' residence, and fellow-commoners were allowed to 'huddle' the exercises demanded of them by the statutes instead of demeaning themselves by public disputation. ('Huddling' consisted of pre-arranged questions and answers which, according to Jebb, enabled a man to read two theses, propound six questions and answer sixteen arguments against them, all in the course of five minutes.) Pensioners were mostly the sons of professional men or the country gentry, while the sizars came from poor families and were generally the most studious body in the university. They received free commons and certain allowances and paid very low fees, but although they no longer served as cleaners or waited at table as they had in the Middle Ages they were ignored socially and treated with disdain by most of the more wealthy students.

Besides the official grades undergraduates were classified unofficially as 'reading' or 'non-reading' men, the latter being known at one period as 'varmints'. The reading men worked seriously with their tutor or private coach for an honours degree, while the others employed their time mainly in social activities, dissipation and fights with the townspeople, but all were obliged to attend a minimum number of college lectures. Heavy drinking was common among the non-reading men and was seldom regarded as an offence by the college authorities. Drunkenness and gambling were the besetting sins of all classes of English society in the eighteenth century, and morals within the university were in general no better than those outside it. Some of the reading men worked regularly and continuously for ten hours a day, took little exercise and kept mainly to their rooms. Most of their studies were in the fields of natural philosophy, which was largely mathematical, and moral philosophy, which included religious knowledge, and these were the only subjects covered by the Senate House Examination for the B.A. degree. However, Latin and Greek authors were read in all the colleges, and the Chancellor's gold medals for classical studies were highly prized.

In the absence of any university teaching staff except the professors, who had no responsibility for the undergraduates, preparation for the degree examinations was entirely in the hands of the colleges. The courses of study were directed by the tutors, who gave most of the college lectures. Trinity had two tutors, one directing the studies in natural philosophy and the other in moral philosophy, and there were several assistant tutors. In the smaller colleges there was normally only one tutor, but as the annual intake of freshmen averaged less than ten per college, of whom more than half were probably 'non-reading' men, the classes were not unmanageable. However, men who aspired to high places in the order of merit were almost bound to resort to private coaches, the best of whom charged very high fees. At the other end of the scale crammers injected a little last-minute knowledge into men who wanted a degree but had not seen fit to work for it.

The exercises in the Schools, which were held during the third year of residence, had changed little since the thirteenth century. They consisted of disputations in Latin—usually a very peculiar kind of Latin—between two candidates who faced each other from rostra on opposite sides of a room under the supervision of the moderators. The best men might have to keep as many as eight of these 'acts', and they provided battles of wits which attracted large audiences and brought great kudos to those who distinguished themselves in argument. The weaker men got off with one or two full disputations and were allowed to 'huddle' the remainder, as their lack of scholastic ability was already known to their tutors. By the end of the eighteenth century the main purpose of the exercises, apart from their educational value—which was still highly esteemed when they were taken seriously—was to put the men thought worthy of the degree into eight classes in preparation for the examination in the Senate House. Men in the first two classes were the 'wranglers', those in the next four the 'senior and junior optimes' and those in the last two οἱ πολλοι or 'the poll men', who were to be awarded the ordinary degree without honours. The classification was carried out by the moderators in consultation with the college tutors, and once it was agreed it was impossible for a man to move from one group to another. Candidates were placed in order of merit within the three groups on the results of the Senate House Examination, which by the end of the century was recognized as the main test of ability before admission to the degree. Later (towards the middle of the nineteenth century) moral philosophy was hived off and it became known as the Mathematical Tripos. It was the direct ancestor of all subsequent examinations for the honours degree, and is therefore worth considering in some detail.

At the end of the eighteenth century the examination was held in January and lasted for four days, three being devoted to mathematics and one to logic and moral philosophy. Candidates entered the Senate House at 8 a.m. and took their places at three long tables in accordance with the class lists, which had been agreed between the colleges and the Examiners and were now on view for the first time. The three tables were for the wranglers, senior and junior optimes and poll-men respectively and each was presided over by two moderators or examiners appointed by the Senate. Pens, ink and paper were provided. The questions were dictated one at a time by the examiners at each table, a new one being read out as soon as the first man at the table was seen to have completed his answer. According to the *University Calendar* of 1802, which gives a full account of both the exercises and the examination,

> The examiners are not seated, but keep moving round the tables, both to judge how matters proceed and to deliver their questions at proper intervals. The examination, which embraces arithmetic, algebra, fluxions, the doctrine of infinitesimals and increments, geometry, trigonometry, mechanics, hydrostatics, optics and astronomy, in all their various gradations, is varied according to circumstances: no one can anticipate a question, for in the course of five minutes he may be dragged from Euclid to Newton, from the humble arithmetic of Bonnycastle to the abstruse analytics of Waring.

While the main body of the candidates was struggling with the dictated questions men in the first two classes were issued with printed papers of problems which they were allowed to take to the window alcoves to answer. The first session lasted from 8 to 9 o'clock, when there was a half hour interval for breakfast. The second session was from 9.30 till 11, the third from 1 until 3 p.m., and the fourth from 3.30 till 5. That completed the day's work for the junior optimes and poll-men, but at seven in the evening the first four classes went up to the Senior Moderator's rooms to solve further problems until nine o'clock when they were dismissed after eight hours of examination. The warmth and comfort of the moderator's rooms and the wine and light refreshment which were often provided there must have been welcome after a long January day in the unheated Senate House. The examination ended at 5 p.m. on the fourth day, when the proctors and Examiners retired to a room under the library to prepare the order of merit. The name of the Senior Wrangler was generally announced at midnight and was loudly acclaimed in a crowded Senate House. The man who came last in the honours list was awarded a wooden spoon, which by the end of the nineteenth century was no ordinary spoon but a maltster's shovel, painted in colours and carried by the recipient over his shoulder.

All who passed the Senate House Examination were sure of their degree, but they had to submit to one more medieval rite in order to comply with the statutes. As soon as the results were announced they were admitted by the Vice-Chancellor 'ad respondendum quaestioni' which meant answering a question put to them in the Schools by the praelector of their college. Formerly this had been a serious viva voce test, but it had long since been reduced to a mere pretence. The praelector asked, 'Quid est nomen?' and the questionist replied 'Nescio'. He was then considered to be fully qualified for the degree, but was not actually admitted until, on the Thursday after mid-Lent Sunday, the proctor or his deputy declared that he had 'fully determined' as a Bachelor of Arts.

THE EARLY PROFESSORSHIPS AND JACKSON'S WILL

The first professorships were established in Cambridge in the sixteenth century. Their principal aim was to attract eminent scholars to the university by offering them security, status and congenial surroundings in which to pursue their studies. In return they were expected to devote themselves to the advancement of their subject, to publish the results of their researches and in some cases to give a specified number of public lectures each year. Most of the early professors carried out their duties conscientiously and some of them brought great renown to the university, but their successors shared in the general decline of academic standards which reached its lowest point about 1760. Discouraged by the lack of interest shown in their work and the scanty attendance at their lectures most of them ceased to lecture at all. Their offices came to be regarded as sinecures, and men were sometimes elected who had not even the rudiments of knowledge of the subjects on which they were supposed to be authorities. After 1760, however, a revival took place, and among the more active professors were those who held chairs in mathematics and the natural sciences. This was in accord with the spirit of the time. The late eighteenth century was an age of invention, and science had become a recognized leisure-time occupation for the more intelligent country gentlemen and wealthy townsmen. Philosophical clubs and societies were established in various parts of the country, scientific papers were written and read, and lectures, often illustrated by demonstrations or experiments, were delivered to fashionable and enthusiastic audiences. It was in this favourable atmosphere that the teaching of applied science developed and that the study of engineering was introduced into the university.

The Lady Margaret, mother of Henry VII, founded the first professorship in Cambridge in 1502. It was in Divinity, and is still known by her name. Nearly forty years later her grandson Henry VIII established the Regius Professorships of Divinity, Greek, Hebrew, Law and Physic. Over a century passed and in 1663 Henry Lucas endowed the first chair of mathematics. By the statutes governing his appointment the Lucasian Professor was required to lecture once a week during term and to be accessible to students who wished to consult him. To ensure that he concentrated on these duties he was forbidden to hold a church living and was compelled to reside in the university. In compensation he was permitted to hold a college fellowship, and notwithstanding any college statutes to the contrary he was not obliged to take Holy Orders. It was these concessions which enabled Isaac Newton to hold the chair as a layman and to retain his Trinity fellowship after election.

Professorships in Arabic, Moral Philosophy and Music were established before the end of the seventeenth century. In 1703 the university created a Professorship of Chemistry for the benefit of John Francis Vigani of Verona, who, although holding no official position, had been lecturing to members of the university on that subject for twenty years. After his appointment a lumber room in Trinity College was fitted up as a laboratory for him, but no formal duties were attached to the office and no stipend was allotted to it. One of Vigani's successors in the chair, who was also vicar of St Andrew's and rector of Landbeach, added to his income by running a business as a dispensing chemist. William Farish, whose career will be described in the next chapter, was holding this professorship when he started the first real course of lectures on engineering in 1794.

In 1704 a Professorship of Astronomy and Experimental Philosophy (which would now be called Physics) was founded by Thomas Plume, archdeacon of Rochester, who left nineteen hundred pounds for the maintenance of the professor, the erection and equipment of an observatory and the provision of a residence in close proximity to it. The Master and Seniors of Trinity covenanted with the trustees to allocate rooms for the professor in the Great Gate and to allow an observatory to be built for him on the top of it. The first Plumian Professor was Roger Cotes, a disciple of Isaac Newton and editor and part author of the second edition of the *Principia.* He died at the early age of thirty-four, and was succeeded by Robert Smith, the founder of the Smith's prizes, who held the chair for forty-four years. One of the greatest of his successors was George Biddell Airy, afterwards Astronomer Royal, who lectured on the construction of domes and suspension bridges as well as on astronomy and mathematics.

Seven more chairs were founded before the close of the eighteenth century—in Anatomy, Modern History, Botany, Geology, Astronomy and Geometry (the Lowndean), Revealed Religion and finally, in 1782, the Jacksonian Professorship of Natural Experimental Philosophy. The Lowndean Professorship was more nearly concerned with subjects required for the Senate House Examination than most of the others, and some early holders of the chair delivered lecture courses which were intended primarily for undergraduates. The Jacksonian Professorship occupies the front rank in the pre-history of the Engineering Department as, for reasons connected with the principal interests of the first three holders, it became the direct forerunner of the Professorship of Mechanism and Applied Mechanics.

The Reverend Richard Jackson of Torrington in Herefordshire, a former fellow of Trinity College, died in 1782 leaving one-fifth of the income from his estate to the chief gardener of the university physic garden and the remainder to endow a professorship. His will laid down in great detail the qualifications required of the holder of this office, the duties which he was to perform and the object which he was to pursue in his researches. He was to read publicly

a course of Lectures on Anatomy, Animal Economy, Chemistry, Botany, Agriculture or the Materia Medica...together with the proper Dissections, Analyses, and experiments upon the subjects of the same, according to their natures, as may compose a due series of experimental lectures and such practice as may truly thence result for confirmation of what is known at present, or for making further discoveries in any branch thereof as may best tend to set forth the Glory of Almighty God, and promote the welfare of mankind...and here it is my serious request that the said Lecturer will in this Disquisition have an eye more particularly to that opprobrium medicorum called the Gout, both in getting a better history of the disorder itself, and the symptoms, preceding, attending and following it, than is extant at present: also of the best method of procrastinating the fits from coming by the use of diet and the other non-naturals: and in finding a medicine that may cure it (of which I have no doubt from the goodness of God that a discovery may be made): and my further request to the Lecturer or Professor is, that herein he will adhere to the plain facts, both in the history and narrative of experiments without adding any hypothesis (unless after the manner that is done by Isaac Newton at the end of his Optics).

An annexe, described as a 'sketch of the form of Lectures in Practical Philosophy which I would have followed by my Professor or Lecturer in the University of Cambridge' contains the following further instructions:

As my design in founding this Lecture is the promotion of real and useful knowledge my opinion is, that this cannot be done to any good purpose by mere reading a Discourse or Disquisition or Essay, but showing or doing something in the way of experiment upon the subject undertaken to be treated...

The field before us is immense, and unless circumscribed by some bounds, and limited by some rules, more likely to puzzle or overwhelm the poor faculties of our minds than to inform and enlarge that information which experiments may bestow, so that among the many objects of enquiry, as my intention is to direct Lectures to the point which may be most use to mankind I recommend an inquiry into those things which we cannot be without, and upon which not only our well-being but our very existence in this world depends. I mean in the first place, the properties of air and water, heat and cold, and what has a considerable share in the effect of all these, the powers of electricity...

So that in short my intention in founding this practical Professorship of Physiology is, that the Reader shall in public experimentally exhibit to young persons the most easy and natural, and best approved method of discovering and demonstrating the medical or nutritive, or domestic properties of bodies of the Mineral and Vegetable Kingdoms, and make the object of his private experiments and disquisitions their more hidden and concealed properties, but to publish them for private good as soon as discovered, and to make those several parts of all, which have been looked upon as useful in gouty cases, the first objects of his enquiry; and see which are useful and where in they are so, and which are worthless, so as to throw them aside.[1]

No Jacksonian Professor has yet discovered a cure for the gout, but the Founder's wish that real and useful knowledge should be sought by the method of experiment had far-reaching consequences, some of which will be described in the next chapter.

THE NINETEENTH-CENTURY REFORMS

The end of the Napoleonic Wars in 1815 led to a general expansion of university education throughout western Europe. In Cambridge matriculations began to rise even before the fighting ended, and by 1820 the undergraduate population was nearly three times as great as it had been at the outbreak of the French Revolution. The increase in numbers was accompanied by a demand for a wider choice of subjects for the honours candidates and a better general education for the poll-men. Some of the reformers wanted an examination to ensure a minimum standard on entry, but this was opposed by the colleges not merely on account of the income they derived from rich young men, who might thereby have been excluded, but because they believed that three years in the university had educational value even for the non-reading men, and should not be denied to those who would inherit positions of responsibility in later life.

The year 1822 marked the beginning of a period of reform which in the space of sixty years transformed the educational system at Cambridge. The first event of importance was the institution in March of a Previous Examination, later known as the 'Little-go', which was compulsory for all degree candidates in their fifth term of residence. It contained elementary papers on the Greek Testament, Paley's *Evidences of Christianity* and a prescribed part of a Greek and Latin author. Its original object was to give the poll-men something to work for other

[1] Quoted in the *Reporter*, 1874–5, p. 253.

than mathematics, but it eventually became the university entrance examination. In May a voluntary examination in classics was approved to be held annually in the Lent Term. It became known as the Classical Tripos, but until 1849 it was open only to those who had obtained honours in Mathematics, and it did not qualify for a degree until 1860. Later in the year papers on Latin and Greek were set in the Senate House Examination for classes which did not take the problem papers. This had the effect of separating the examination for the ordinary degree from that for honours and made it possible for the colleges to organize their teaching on two levels.

In 1827 printed papers were used for the first time in the Senate House Examination, and five years later Divinity, Mechanics and Hydrostatics were added to Mathematics and Classics in the examination for the ordinary degree. Preparation in all these subjects ensured for the poll-men a more general education than formerly and paved the way for the teaching of applied science later. The medieval exercises, which had long ceased to fulfil any useful purpose for most of the candidates, were held for the last time in 1839.

The further extension of undergraduate studies was hindered for many years by the conviction that mathematics, 'the honour and glory of the University', was the best of all possible intellectual disciplines. The natural sciences, which were gaining a foothold in European and Scottish universities, only fitted into the Cambridge pattern when they could be treated mathematically, and the moral sciences, languages, literature and history, which could not be so treated, were considered to have little educational value. The chief protagonist of the mathematical discipline was William Whewell, Master of Trinity from 1841 to 1866, who towered above his contemporaries in much the same way as his predecessor Richard Bentley had done, though without his violence and lack of scruple.

The reforms of the twenties and thirties did much to raise the standard of the degree examinations, but they had some unexpected repercussions which were educationally deplorable. Competition for the top places became fiercer than ever, and, as the papers were much more difficult than formerly, would-be wranglers had to restrict their studies as rigorously as possible to the syllabus for the tripos. They regarded the Previous Examination as a nuisance which compelled them to divert attention from their main objective during five precious terms of their residence. Men whose abilities were not mathematical also suffered. Formerly they had been able to read widely in the classics and still master enough mathematics to achieve a degree with honours, but now they had to concentrate their attention on advanced mathematics at the expense of subjects more suited to their talents. Even the poll-men failed to benefit from the changes in the way that had been expected, as they felt obliged to 'forego the luxury of mental improvement in order to make sure of passing their examinations'.[1] Another effect of the raising of standards was that the tutors, particularly in the smaller colleges, proved unable to carry the additional teaching load imposed on them. The honours candidates were therefore compelled to resort to expensive coaches, while the poll-men relied on crammers to achieve a last-minute miracle in the examination.

One of the difficulties in the path of those who wished to extend the scope of undergraduate teaching was thus the lack of any university teaching staff. Whewell conceived the idea of employing the professors in subjects which might be considered suitable for degree examinations, and in a book entitled *Of a Liberal Education*[2] he suggested that all candidates for the ordinary degree should be required to attend at least one course of approved professorial lectures. He also advocated the establishment of a tripos examination in those branches of natural science in which chairs already existed, and in 1842, as Vice-Chancellor, secured the appointment of a syndicate to consider the possibility of relating the professorial lectures on mathematics and classics to the triposes in those subjects. The syndicate devised a scheme for putting his ideas into practice, but the colleges resented the proposal that the university should encroach on their teaching preserve and the report was rejected in the Non-Regent House.

Another cautious step forward was, however, taken in 1846, when the Senate House Examination for the honours men, now known as the Mathematical Tripos, was divided into two parts. Part I was comparatively simple, and was taken by all candidates. Eight days

[1] D. A. Winstanley, *Early Victorian Cambridge* (Cambridge, 1940), p. 178.
[2] Published in three editions in 1845, 1850 and 1852.

later a second examination of a much more advanced nature was held for those who were considered to have reached honours standard in Part I and on this was based the final order of merit of the wranglers and senior and junior optimes.

By this time many friends of the university were becoming convinced that real reform could never come from within, and as a first step towards compulsory modernization by Act of Parliament were urging the appointment of a Royal Commission of Inquiry. The prospect of this was abhorrent to most members of the Senate, but the threat undoubtedly instilled a spirit of urgency into the university and acted as a spur to the reformers. The sudden death of the Chancellor, the Duke of Northumberland, enabled Whewell to enlist a powerful ally in the person of Prince Albert, who was elected to the vacant post in February 1847. Throughout his period of office, which lasted until his death in 1861, Prince Albert's chief adviser was Dr Henry Philpott, Master of St Catharine's, a wise, broad-minded man who was less tied than Whewell to the maintenance of the mathematical domination. At Philpott's suggestion a syndicate, of which both he and Whewell were members, was appointed to study the teaching and examination systems and make suggestions for reforming them.

The syndicate's report appeared in April 1848, and its recommendations were submitted to the Senate in the form of five Graces which were to be voted on after the Long Vacation. They included an improved version of Whewell's scheme for bringing the professors into undergraduate teaching for the ordinary degree examination, and also proposed the establishment of two new triposes, in the natural and moral sciences, for the honours men. The Natural Sciences Tripos was to include papers on anatomy, physiology, chemistry and geology, and the Moral Sciences Tripos papers on philosophy, political economy, modern history and jurisprudence. As in the Classical Tripos at that time success would carry honours but would not qualify for any degree. No additions to the university teaching staff were suggested and it was unrealistically assumed that teaching in all these subjects would be given by holders of existing chairs or by the college tutors. A further proposal, which had far-reaching consequences, was that the Lucasian, Plumian, Lowndean and Jacksonian Professors, together with the

Moderators and Examiners for the Mathematical Tripos, should form a Board of Mathematical Studies to consult together 'on all matters relating to the actual state of mathematical studies and examinations in the University' and to prepare a report on them annually for publication by the Vice-Chancellor. To the surprise of both supporters and opponents of these revolutionary proposals all five Graces were approved by substantial majorities when they were submitted to the Senate on 31 October. This result was acclaimed by the press as a victory for reason and common sense, and most of the credit for it was rightly given to Prince Albert.

Reform from within had, however, been too long delayed, and during the Long Vacation of 1848, while the report of the Teaching Syndicate was under consideration, a memorial signed by over two hundred people, more than half of whom were Cambridge graduates, was presented to the prime minister, Lord John Russell. It declared that the Universities of Oxford and Cambridge had failed in their duty to promote the advancement of learning and that, as they were incapable of reforming themselves, the government should appoint a Royal Commission of Inquiry as a first step towards their compulsory modernization by Act of Parliament. Lord John Russell had great respect for Prince Albert, and would willingly have allowed him more time to put his own house in order, but pressure for parliamentary intervention continued to increase and eventually became irresistible. In April 1850 he accepted a private motion asking for the appointment of a Royal Commission to inquire into the state of Oxford and Cambridge Universities and Trinity College, Dublin, and the Queen, as a constitutional monarch, accepted his advice.

The news aroused great indignation in Cambridge, but Albert pleaded earnestly for patience and restraint, and as time passed passions cooled somewhat and the tension eased. The nomination of the commissioners did much to allay anxiety as they were all Cambridge men whose judgement and integrity inspired confidence in the university, and although the Masters of Jesus and Clare refused to co-operate in any way and Whewell protested that the commissioners had no legal right to call for papers opposition gradually died down and a mass of evidence was collected and studied.

The report, a massive document, was published in November 1852. It put forward a comprehensive plan for the complete reorganization of the educational system of the university. New triposes were proposed in Theology, Modern Languages combined with History, and Engineering, and Boards of Study, on the mathematical model, were to be set up for Classics, Theology, Law, Moral and Natural Sciences and Medicine. The examination for the ordinary degree came in for severe criticism, and the commissioners recommended that instead of one examination alternatives should be offered in all subjects qualifying for honours, but of a lower standard than the triposes. Ten new chairs were suggested, including one in Practical Engineering and one in Descriptive Geometry. College tutors were to be responsible for preparing their undergraduates for the Previous Examination, but candidates for honours and for the final examination for the ordinary degree were to be taught by professors and a new staff of university lecturers. Money for the stipends and for the running expenses of the lecture rooms and museums which would be required for university teaching was to be obtained mainly by a levy on the income of the colleges.

The threat of this levy and the suggestion that the university should take over so much of the teaching aroused bitter indignation in the colleges. However, two new syndicates were immediately appointed to consider the commissioners' proposals. One was to report on the feasibility of establishing a staff of 'Public Professors and Public Lecturers' and the other to consider how additional lecture rooms and museums could be provided. The first soon ran into trouble, and in March 1854 Philpott admitted that 'there are so many members of the syndicate skilful in raising objections, and indisposed to help in making progress that it is difficult to say when the report may be published'. The other syndicate recommended that a building containing a large lecture theatre, lecture rooms and museums should be built on the site of the old Botanic Garden, but when it was found that the cost would be about £23,000 the scheme was abandoned.

Meanwhile the colleges did little to reform themselves, and as their co-operation was essential if the commissioners' plan for the university was to be implemented the demand for a Statutory Commission grew more insistent. At last Lord John Russell informed Parliament that in his opinion Cambridge needed a more representative form of government, that sinecure fellowships should be abolished and a much greater proportion of college revenues devoted to teaching. A bill to reform Oxford's statutes was introduced in March 1854 and notice was given that Cambridge would soon receive the same treatment. Philpott and others, wishing to keep the initiative as far as possible in the hands of the university, drafted proposals for a bill on similar lines and Albert passed it to the prime minister. It was accepted by Parliament after slight amendment and became law in July 1856 as the Cambridge University Act. Its most important provisions were the abolition of religious tests for all degrees except those in Divinity (though the Senate was still restricted to members of the Church of England), the replacement of the Caput[1] by a Council of the Senate and the appointment of eight commissioners, all Cambridge men, to revise the statutes of the university and the colleges.

The Statutory Commission held its first meeting on 27 October 1856 and the Council of the Senate was elected in November. Its first duty was to collaborate with the commissioners in the drafting of the new statutes, and under the chairmanship of Philpott, who at this critical time was fortunately elected to a second period of office as Vice-Chancellor, it carried out its task with remarkable success. A new Code of Statutes was hammered out, and received the formal assent of the Queen in Parliament on 31 July 1858.

The colleges, however, proved much more stubborn, and when the commissioners issued their final report in 1861 they had to admit failure on many important issues. As only Trinity, Peterhouse and Christ's were willing to contribute to a central education fund the proposal for a levy on the colleges was abandoned. 'It is to the failure of our attempts to derive assistance from this source', wrote the commissioners, 'that we must mainly ascribe the imperfect manner in which we have been able to remedy the existing defects in the Professorial arrangements of the University.'

Like their predecessors in the Royal Commission the Statutory Commissioners advocated the establishment of ten new chairs, but their list

[1] A body elected by Heads of Houses, doctors, and the two tellers in the Non-Regent House from a list prepared by the Vice-Chancellor and the proctors. Its unanimous approval was required before a Grace could be submitted to the Senate.

showed several important differences from that of 1852. From the point of view of technology the most important was the omission of Engineering and Descriptive Geometry and the substitution of Electricity and Magnetism. The reasons for this will be discussed in the next section.

The new statutes contained no regulations for examinations, which in future were to be put into ordinances drawn up by the university itself. A body which became known as the Teaching Syndicate began work on their revision, and in 1860 the Senate agreed that men who passed with honours in any tripos would be admitted to the B.A. degree provided they had been in residence for three years. In 1865 the regulations for the ordinary degree were completely recast. Candidates were now required to take a General Examination in the Easter Term of their second year and a Special Examination in any one of five subjects a year later. One of these 'Specials' was in Mechanism and Applied Science, and it was thus that certain aspects of engineering became for the first time the subject of a degree examination in Cambridge.

A matter of overriding importance with which the Statutory Commissioners had felt unable to deal was brought before Parliament in 1870, when a bill for abolishing all religious tests except for Heads of Houses and candidates for Divinity degrees was introduced as a government measure. It became law in the following year. The admission of Roman Catholics, dissenters and free-thinkers to college fellowships and university teaching posts changed the whole character of the university. A Church of England preserve became a national institution of learning in a sense which had not been possible since the Reformation. A whole reservoir of intellectual talent which had been barred from Cambridge solely on religious grounds now became available, and the stimulus which this gave to teaching and research soon became apparent. One of the first to benefit from the abolition of the tests was John Hopkinson, whose influence on the early history of the Engineering Department will be described in due course.

Without a financial levy on the colleges the provision of a staff of university lecturers was impossible, but money was found for a number of new professorships. Between 1851 and 1882 twelve new chairs were founded, including the Cavendish Professorship of Experimental Physics in 1873 and the Professorship of Mechanism and Applied Mechanics two years later. The lecture room and museum block recommended by the syndicate in 1852 was built in 1864 in the old Botanic Garden, which has ever since been known as the New Museums site. No money was available for a laboratory of Experimental Physics, but the Duke of Devonshire, who had succeeded Prince Albert as Chancellor in 1861, built and equipped the Cavendish Laboratory at his own expense and presented it to the university.

Between the middle years of the century and the 1880s the number of matriculations rose from about 400 a year to over 800. Even with the help of the new professors college tutors were unable to carry the teaching load imposed by the examination requirements, and the demand that the colleges should devote a larger proportion of their incomes to teaching gradually became irresistible. In January 1872 a Royal Commission was appointed 'to inquire into the property and income belonging to, administered or enjoyed by the Universities of Oxford and Cambridge and the colleges and halls therein...'. Its report appeared in the summer of 1874 and revealed that many of the colleges possessed incomes far in excess of their existing educational commitments.

The case for a financial levy had been further strengthened when, in 1873, a Royal Commission of Inquiry on scientific education in England reported that out of 350 fellowships in the Cambridge colleges only 120 were held by residents doing educational or administrative work inside the university. In that year a memorial signed by over one hundred people, including twenty-six professors, was presented to Gladstone, the prime minister. It asked

> that fellowships divorced from work at the University should not be held for life; that a permanent professional career should be opened to those engaged in college work by allowing fellows of colleges to marry; that provision should be made for the association of colleges for educational purposes in order to secure more efficient teaching and more leisure for study; and that the pecuniary and other relations between the University and the colleges should be revised.

Statutory Commissioners for both Oxford and Cambridge were appointed in August 1877 with instructions to provide for college contributions to a university education fund, to attach fellowships and other college emoluments to university offices and to review the tenure of fellowships not attached to such offices. In Cambridge the Council of the Senate again collaborated loyally with the commissioners, and in 1882 their task was completed. Under the revised statutes a scale of college contributions to the University Chest was fixed, professorial fellowships were established and fellows were exempted from the requirement to take Holy Orders or to abstain from marriage. A Financial Board was set up to handle the increased university income, and a General Board of Studies to co-ordinate the work of Special Boards in the various fields of teaching and research. The transformation of the university as it had existed at the beginning of the century was complete, and the way was open to further progress and developments on modern lines.

THE GENESIS OF THE ENGINEERING DEPARTMENT

The proposal to establish a School of Engineering at Cambridge was first considered seriously in 1845, but the collapse of the railway boom led to a general depression in the engineering industry and the idea was temporarily abandoned. Chairs of engineering were, however, founded in other colleges and universities in Britain during this period. The subject had been taught in King's College, London, since its foundation in 1831, and a 'Professorship of the Arts of Construction in connection with Civil Engineering' was established there in July 1840. Four weeks later a Regius Professorship of Engineering was established somewhat arbitrarily in the University of Glasgow. University College London, founded in 1828, had, like King's College, London, given engineering courses from the beginning, and it obtained a chair of Civil Engineering in 1841. Trinity College, Dublin, followed suit in the same year. Engineering was also taught at Durham University from its foundation in 1837 and in Queen's College, Belfast, after 1843, though neither yet had a professorship in the subject.

In Cambridge the Royal Commission of 1850 came out strongly in favour of a special course for the education of engineers, and its report of 1852 devoted much attention to the subject. The commissioners described the profession of engineering as one of the most important and lucrative employments in the country, and expressed the view that engineering was a legitimate subject for university study and teaching. Many of the practical and technical details could, no doubt, be taught best in the industry itself, but no amount of practical skill and experience could replace the theoretical knowledge needed in the most advanced applications of the science. The lectures of the Jacksonian Professor Willis, which will be described in the next chapter, were referred to as excellent examples of the union of theory and practice, and the hope was expressed that if a School for the benefit of intending engineers were established in the university he would widen the scope of his courses, which at present were intended mainly as a contribution towards a general education. The commissioners further suggested that Professors of Practical Engineering and Descriptive Geometry should be appointed, and that under the general supervision of Willis and the Board of Mathematical Studies courses should be given on drawing, descriptive geometry, the principles of construction and other subjects required by a student before entering the office of a practising engineer. The resulting course and the examination in it should qualify for the B.A. degree.

In February 1853 the commissioners' suggestions regarding engineering were studied in detail by the Studies Syndicate set up after the publication of the Royal Commission's report to consider the proposals for new triposes and a university teaching staff, and one member is known to have suggested that instead of founding new chairs the Jacksonian Professor should take over the lectures on engineering and the Lowndean those on descriptive geometry. The professors' reactions to this proposal are not recorded, but it is clear from subsequent events that Willis had no wish to undertake the organization of a School of Engineering at Cambridge. He had just completed his work as a Juror of the Great Exhibition and was shortly to become a lecturer at the government's new School of Mines in London, and in addition he was deeply interested in medieval architecture and archaeology. He was certainly not looking for further responsibilities.

One of the few surviving records of the Studies Syndicate is a list of propositions marked 'for the Studies Syndicate only' and dated 10 May 1854.[1] Proposition No. 57 is

[1] University Archives.

That the question whether Civil Engineering should be recognised as a subject of study in the University deserves further consideration: and that a sub-syndicate be appointed to confer with Professor Willis and others on that subject; also to ascertain what course of instruction is adopted in Schools of Civil Engineering; what additional teachers it would be necessary for the University to appoint to carry out the scheme; and in what way the study might be more effectively encouraged by the University; and to report on the subject generally to the Syndicate.

The sub-syndicate was duly appointed, but its membership is unknown and none of its reports are to be found in the University Archives. It is, however, on record[1] that a correspondence was opened with the British universities and colleges which had already established Schools of Engineering and that their reports were so discouraging that the syndicate felt unable to recommend the establishment of such a School in Cambridge. This decision was strengthened because many members of the engineering profession who took articled pupils were said to be hostile to the proposal—a striking contrast to the view expressed by the commissioners 'that the really eminent and well-educated members of that important profession are too well aware of the paramount importance of a scientific knowledge of mechanical principles and their application, in the practice of their art, not to co-operate in the promotion of any attempt to give a proper preparatory education to those who propose pursuing it'.[2]

This was indeed an unfortunate time to consult the other academic schools, for they were all in trouble. At Glasgow the first Regius Professor L. D. B. Gordon, a practising engineer who had worked under the great I. K. Brunel, found it difficult to establish himself in the face of the hostility of his colleagues in the chairs of Mathematics and Natural Philosophy. Pupils were hard to come by, and he is believed to have discontinued his lectures entirely from 1851 to 1854 before resigning in the following year. In London, as already noted, engineering had been taught in University and King's Colleges since their foundation, but both schools were at a low ebb in the mid fifties. University College had not been very fortunate in its first engineering professors and it attracted relatively few students in that subject until

[1] H. Latham, *On the Establishment in Cambridge of a School of Practical Science* (Cambridge, 1859); see below.

[2] Report of the Royal Commission, 1852.

after the Crimean War. King's College got off to a better start, but there too numbers declined during the railway slump and an incipient recovery was checked by the outbreak of the Crimean War in March 1854. It is therefore not surprising that Engineering was dropped from the list of ten new professorships recommended by the Statutory Commissioners in the final report which they issued in 1861.

Two years before this report was published, while it still seemed possible that the Cambridge colleges would contribute to a general education fund and that money would be available for new teachers, lecture rooms and laboratories, Henry Latham, tutor of Trinity Hall, put forward an outline plan for the establishment of a School of Practical Science on continental lines. During the Long Vacation of 1859 he visited the Polytechnic School at Karlsruhe to study the German system of technical education at first hand, and on his return published a pamphlet[1] which is of considerable interest. In it he pointed out that the Cambridge system of teaching theoretical science as part of a general education was an entirely different matter from the teaching of applied science as an education for engineers. In the school which he advocated some of the teachers would be practising engineers (as was the case in Scotland) and it would be controlled by a special Board of Studies on which he hoped that Professors Willis and Stokes, who were in charge of departments in the government School of Mines, would exert the greatest influence. Most of the students would be articled as apprentices before or after starting the course, and as this class of person would be unlikely to know any Greek they would be excused the 'Little-go'. They should, however, be required to pass a special entrance examination if they had not obtained a certificate in the University Local Examinations, and at some time during the course they would have to take the mathematical papers for the ordinary B.A. degree as medical students were obliged to do. They would be matriculated and admitted to colleges like undergraduates, but would not be compelled to keep more than four terms, though they would be encouraged to stay longer. The terms need not be consecutive but could be combined with practical work in their firms on the 'sandwich' system. Lecture courses would be given on (1) Mechanism applied to

[1] Latham, *On the Establishment in Cambridge of a School of Practical Science.*

the Arts, (2) Civil Engineering—two terms, (3) Geology, (4) Mineralogy, (5) Chemistry, (6) Metallurgy, and (7) Field Surveying. Mechanical and Geometrical Drawing might be given throughout the course. Each student would be obliged to attend at least four courses, Civil Engineering counting as two, and to pass an examination in them, after which he would be admitted to the title of Associate in Practical Science or A.P.S. A metallurgical laboratory would be required, but that already possessed by St John's College would answer the purpose until a larger one could be built for the university. A workshop would be needed for the instruction of mechanical students, and a collection of philosophical instruments and apparatus which he described as a 'physical cabinet'. The existing museums of Mineralogy and Geology would be 'ample and excellent'. Besides the professors in the appropriate scientific subjects and the professional engineer already referred to, lecturers in surveying should be appointed. Instruction should be given in the methods of preparing estimates, in getting up plans for parliamentary committees and similar practical matters.

Latham's plan was abandoned when the colleges refused their contributions to the Chest, and no other attempt to found a chair of Engineering was made while Willis lived, but many of his ideas persisted and bore fruit in later years.

Willis died on 28 February 1875, and the need for a new appointment to the Jacksonian Chair drew attention to the fact that lectures on mechanism were difficult to reconcile with the terms of the endowment. The Council of the Senate published a report containing long extracts from Jackson's will and suggested that the electors should give the new professor some guidance regarding the subject of his teaching. In the discussion of the report Latham pointed out that if Mechanism were not to be the new professor's subject other arrangements would have to be made, as candidates for the Special Examination in Applied Science were required to produce a certificate that they had attended such lectures. Another speaker put the dilemma clearly before the Senate:

> If the instruction in Engineering is abolished an existing branch of study, capable of large expansion, will be destroyed; if the Professorship is appropriated to Engineering an act of violence will be done to the founder's will. Either there must be two Professors or one Professor must combine the two works.

A syndicate was appointed to consider the matter, and it reported on 6 May, recommending the establishment of a Professorship of Mechanism and Applied Mechanics which would terminate with the tenure of the first holder unless the university decided otherwise. The professor was to teach and illustrate (1) The Principles of Mechanism, (2) The Theory of Structures, (3) The Theory of Machines and (4) The Steam Engine and other Prime Movers. His scheme of lectures was to be subject to the approval of the Board of Mathematical Studies, of which he would be a member *ex officio*. He was to be chosen by all the members of the Senate on the electoral roll, and his stipend was fixed at £300 a year.

In the discussion on the report the first speaker deplored the omission of the word Engineering from the title of the proposed chair, but this was defended by a member of the syndicate on the grounds that engineering was a vague term 'with no ascertained technical meaning'. He said the syndicate wanted a professor 'who would be useful in case a school of Engineering arose in the University' and it was with this end in view that they had defined the subjects he would be expected to teach.

This somewhat ambiguous statement seemed to mean that the creation of the professorship would not necessarily imply the establishment of a school of Engineering, but the next speaker, who was also a member of the syndicate, explained that they did in fact wish to establish a school of Mechanism which would not only prepare candidates for the Special Examination but would be beneficial to the country at large; they had only excluded such subjects as the construction of railroads, canals, docks and drains which were too practical to be included in a university syllabus. Another speaker criticized the limitations proposed by the syndicate. There were, in his opinion, two methods of establishing an engineering school in a university. One was to attract a man of recognized eminence in the profession to set such a school going by the impetus of his name, but no such person should work full time for a stipend of £300 a year. Another method was to get a good young engineer to build up a school gradually while spending his spare time in the industry, but the establishment of any permanent school was ruled out by limiting the professorship to a single tenure. None of the speakers objected in principle to the creation of a new chair, and the Vice-Chancellor therefore gave notice that a Congregation would be held on 28 October to establish a professorship on the terms laid down in the report.

This announcement was followed by a number of flysheets and letters to the press, some for and some against the scheme. The chief opponent was Robert Phelps, Master of Sidney Sussex, who is described by Winstanley as by far the most active and extreme of the reactionaries. His flysheet denounced the establishment of a Professorship of Mechanism on the grounds that it was not wanted and that the charge on the Chest would be an unjustifiable extravagance and embarrassment. In letters to the *Cambridge Chronicle* he inveighed against the whole system of professorial lectures, which he considered a most inefficient form of teaching. A professor's function was to carry out original research, not to lecture to undergraduates. There were plenty of able mathematicians in Cambridge competent to teach geometry and mechanics and to explain the principles of mechanism and engineering. If any special apparatus were required it should be paid for by the tutor or the pupils, not by the University. It was absolutely ludicrous to think of setting up anything like a practical school of mechanism and engineering in Cambridge, and for anything short of that everything needed was already available. On the subject of the professor himself—a man of special genius 'who would inspire ideas of substantial progress in those important branches of knowledge'—he was particularly scathing.

Well, Sir, I really think I need only express my amazement that any sensible man can be so weak as to believe that, in these days, when the possession of anything like such a genius would insure to its distinguished possessor an enormous fortune in any country in the world, the offer of £500—or even £5000—a year would induce such a man to give his 'undivided attention and energies' to the real interests of those truly important departments of practical science as a professor in the University Town of Cambridge...To believe that the University will be able to secure for a professor—now, in these days—such a man as I have described, and as would alone be worthy of the name, in connection with those practical sciences, requires an amount of credulity to which I hope ever to be a stranger.

The Senate, however, did not agree with this point of view and the Grace for the professorship was approved by 74 votes to 36. In a further letter to the *Cambridge Chronicle* published on 30 October, Dr Phelps reported

> with profound regret and humiliation as a member of the University that the Senate has voted the establishment of a Professorship of Mechanism and Applied Mechanics, with a stipend of £300 a year out of the University Chest. As I, of course, expected, the whole force of University Radicalism and the corps Professorial were in favour of the Grace...However, I am happy in having publicly protested...

2 PROFESSORIAL LECTURES ON ENGINEERING AND APPLIED SCIENCE, 1783–1875

ISAAC MILNER, 1750–1820

An account has already been given of the founding of the Jacksonian Professorship of Natural Experimental Philosophy through which the study of Engineering was introduced into the university. The first holder of the office was Isaac Milner, who was born in Leeds on 11 January 1750, the third and youngest son of very poor parents. The father, who had been brought up as a Quaker, was a conscientious and hard-working man, and it was his ambition to give his children a better education than he himself had enjoyed. He could do little in this respect for his eldest son, who was put to work at an early age, but by exercising the utmost parsimony he was able to send the two younger boys to the local grammar school. However, he died in 1760, leaving their education unfinished and his widow penniless.

Joseph, the second son, who was then seventeen years old, had attracted the favourable notice of his headmaster and through his good offices obtained a sizarship at Catharine Hall—now St Catharine's College—at Cambridge. He had no natural aptitude for mathematics, but by sheer industry and determination, aided by a prodigious memory, he graduated as third Senior Optime in 1766. His real interests lay in the classics and in that field he distinguished himself by winning the second of the Chancellor's gold medals in the face of severe competition. On leaving Cambridge he took Holy Orders, and after two years as an assistant in a small school near Tadcaster was appointed headmaster of Hull Grammar School and afternoon lecturer in the principal church in that town. He now had an income of upwards of two hundred pounds a year, and his first thought was for the less fortunate members of his family. He installed his mother as his housekeeper, a post which she filled 'with great cheerfulness and activity' for over twenty years, and made himself responsible for the education of the two orphan sons of his eldest brother, who had recently died.

Isaac had been apprenticed to a woollen manufacturer when his father died, but while working in the mill he had done his utmost to continue his education by reading and private study. Joseph determined to apply for his appointment as usher in the Grammar school at Hull and enlisted the aid of his local vicar, the Rev. Myles Atkinson. Atkinson went to Leeds to inquire into Isaac's qualifications for the post, and on arrival at the mill found him seated at the loom with copies of Tacitus and a Greek author open beside him. The record does not state whether the young man had received prior warning of the visit. The interview went off well, and the mill owner good-naturedly agreed to release him from his engagement, dismissing him with the words, 'Isaac, lad, thou art off!'

He embarked on his duties as usher with the utmost enthusiasm. Unlike Joseph he had a great flair for mathematics, and as well as giving general instruction to the younger boys he occasionally helped with the teaching of algebra in the upper school. His all-round intellectual ability was outstanding, and in 1770 he secured admittance to Queens' College, Cambridge, as a sizar. In 1774 he graduated as Senior Wrangler and first Smith's prize man, so far ahead of his rivals that he was awarded the distinction of 'incomparabilis', and with these triumphs to his credit his academic future was assured. He was ordained deacon in 1775 and two years later he was appointed College Tutor, in which capacity he soon acquired a very high reputation. He proved to be a first-rate teacher, and the mathematical papers which he set were considered so remarkable for their neatness and elegance that a student from another college tried to bribe his bed-maker to 'borrow' a set for him to copy.

After graduating as M.A. he studied chemistry under Professor Watson and later lectured as deputy for Watson's successor Pennington. A serious injury to his lungs, caused by inhaling a poisonous gas while conducting a chemical experiment, may have caused a permanent weakness, and in spite of a robust appearance he was subject to periodic nervous and physical illnesses which completely prostrated him. He took priest's orders in 1777 and in the following year was presented by his college to the rectory of St Botolph's in Trumpington Street, where a curate carried out the work of the parish for him. Although a deeply religious man and a great preacher he always pleaded ill-health as a reason for not binding himself to conduct regular church services, and was sometimes suspected of evading duties which were not congenial to him.

In addition to his work on chemistry he continued his mathematical studies and in the spring of 1777 communicated to the Royal Society a paper on algebraic equations which was read for him by the Plumian Professor. A second paper, on the communication of motion by impact and gravity, followed in the next year and a third on the precession of the equinoxes in 1779. In 1780, at the age of thirty, he was elected to a fellowship in the Society. He acted as moderator in that year and again in 1783 and was proctor in 1781–2, when he is said to have escaped unpopularity by the firm but good-humoured manner in which he discharged his duties.

He was now a well-known and respected figure in the university, and it caused no surprise when in 1783 he was elected to the newly established Jacksonian Professorship of Natural Experimental Philosophy. Unlike most of the eighteenth-century professors he took the founder's instructions very much to heart, and from 1786 to 1792, apart from interruptions due to illness, he read alternate courses of lectures on chemistry and experimental philosophy.

Experimental philosophy was an eighteenth-century expression with a meaning that was never clearly defined. It included what would now be called physics, mechanics, metallurgy, chemistry and applied science in general, and in Jackson's will it was given an even wider scope. The suggestion that lectures on these subjects should be illustrated by experiments or demonstrations was not a new one. George Atwood of Trinity, who was third wrangler and first Smith's prize man in 1769, lectured on 'The Principles of Natural Philosophy' in the observatory over the Great Gate during the seventies, and his notes were published in booklet form in 1776. They listed 5 experiments in mechanics, 46 in hydrostatics, 35 in electricity, 10 in magnetism and 54 in optics as well as notes and headings on astronomy.

Milner adopted an equally practical approach to his subject, and his lecture notes are of particular interest in the history of engineering at Cambridge as they contain the first notice of instruction in the theory of the steam engine as well as problems involving air pumps and other mechanisms. He was assisted at his lectures by a German named Hoffman, who illustrated them by a series of practical demonstrations which kept the audience in a constant state of interest and entertainment. Milner possessed a fund of high spirits and a boisterous humour which appealed to young men, and his courses became extremely popular. The chemistry lectures were always well attended, but those on natural philosophy were described as amusing rather than instructive. Milner himself took them all seriously, and in a letter written in 1787 he described his daily routine.

In college I lecture from eight to ten in the morning—from that time till four in the afternoon I am absolutely so engaged that I can scarcely steal half an hour from preparing my lectures to dine. At half past five I get my coffee, go to chapel, and then lie down for an hour—then I rise, take my milk—look out various articles, and make notes on natural history, etc., for the succeeding day. This coming every day keeps me on such a continual stretch that I am often very much done up with fatigue; and if Mr Metcalfe of Christ's College did not assist me I should not be able to get through.

It should, perhaps, be explained that this letter was written to discourage an acquaintance who wished to pay him a visit but whom he was not anxious to entertain.

Milner's interest in practical mechanics soon became widely known. By 1784 he was spending considerable sums of money on equipment for his lectures and private researches, and in a letter to the reformer William Wilberforce, who had been his pupil at Hull Grammar School, he referred to bills for electrical apparatus, air pumps, furnaces and crucibles. In 1788 he was elected President of his college, and when he moved into the Lodge he fitted up a large room as a workshop and laboratory. He installed lathes, furnaces, work-benches, grind-stones, bellows, blow-pipes, electrostatic appliances, etc., and there, either alone or in collaboration with friends with similar interests, carried out chemical, electrical and mechanical experiments. In a letter written in 1791 he said, 'I have been a great dabbler in air-pumps, and have spent a great deal of money on them. I have now one by me which cost £60 and upwards...they are my hobby-horses, and there is no saying what lengths a man will not go, to gratify himself in such cases.'

As President of Queens' he was responsible for many reforms, some of which brought him into conflict with his fellows, but his zeal and energy had on the whole a beneficial and lasting effect on the college. He was a man of strong opinions, and a bitter opponent of the revolutionary tenets of Jacobinism, which he did his utmost to extirpate from the university. He was strongly opposed to the abolition of the religious tests, which debarred Roman Catholics and Nonconformists from university office, but supported Wilberforce in his campaign for the abolition of the slave trade. He became the leader of the Tory Evangelical party in the university and in his last years acquired immense authority. He had been admitted to the degree of Bachelor of Divinity in 1786, when his 'act' was acclaimed to be one of the finest on record. In 1791 he was nominated dean of Carlisle, and in the following year became a Doctor of Divinity. He was Vice-Chancellor in 1792, but he was at this time a very sick man and was obliged to obtain exemption from some of his duties—including attendance at the services in Great St Mary's. He resigned from the Jacksonian Chair and did no more public lecturing. However, his health improved and in 1798 he was elected to the Lucasian Professorship. In a letter to Wilberforce he declared that the duties of the new office would give him no trouble at all, though he meant to be as efficient as possible in discharging them. They included the examination of candidates for the Smith's prizes, a task at which he excelled, and the general encouragement of the study of mathematics in the university.

From this time until his death he spent about one-third of his time in Carlisle and the remainder mainly in the President's lodge at Queens'. He became more and more devout and was always willing to preach in the cause of the evangelical revival, but in spite of all his religious and college activities he continued to devote much of his leisure to mechanical hobbies and his reputation as an engineer eventually attracted the notice of the government. In 1801 he was asked to prepare for official use a memorial about the construction of a new bridge over the Thames. It seems probable that he owed this commission to Wilberforce, to whom he wrote a long letter explaining his proposals. He claimed to have given the matter a great deal of attention and to have consulted many authors. He was convinced that a conference between the practical engineers and the theorists was essential, and he recommended Professor Farish, a former pupil and friend of many years standing whose career will be described in the next section, as the person whose mathematical and mechanical knowledge might be most useful in this respect. 'I am', he asserted, 'most positive that there is nobody here equal to him or to be compared.' It is not known whether either Milner or Farish had any further connection with the design of the Thames bridge, but the incident is of interest as it may well be the earliest authenticated instance of the British government seeking advice on a matter of practical engineering from a university professor.

Milner died on 1 April 1820, and with his going the university lost one of its most colourful and controversial personalities. One who knew him well wrote of him,

> He had looked into innumerable books, and dipped into most subjects, whether of vulgar or learned enquiry, and talked with shrewdness, animation and intrepidity on them all. Whatever the company or whatever the theme his sonorous voice predominated over all other voices, even as his lofty stature, vast girth and superincumbent wig defied all competitors.

He was a man to be reckoned with, and there can be no doubt that his interest in engineering problems did much to make the subject respectable in Cambridge and opened the door to further and more important developments in the years which followed.

WILLIAM FARISH, 1759–1837

William Farish was the first man to teach the construction and use of machines as a subject in its own right instead of merely using mechanisms as examples to illustrate the principles of theoretical physics or applied mathematics. The lectures which he started in 1795 led directly to the more specialized courses of Willis and the acceptance in 1865 of engineering as a subject for a university examination and a qualification for a Cambridge degree.

He was born in Carlisle in 1759, the third son of the Rev. James Farish, vicar of Stanwix and lecturer and minor canon in Carlisle cathedral, a man of taste and culture, a linguist, botanist and musician and an amateur in the scientific discoveries of the day. His mother came from a well-connected local family named Gilpin. There were three other sons and four daughters. William attended the local grammar school, but received the most important part of his education at home from his father. He was a clever child, and on 23 March 1774, at the early age of sixteen, he was entered as a sizar at Magdalene College. He matriculated in the Easter Term 1775 and in November of the following year he was elected to a college exhibition. He proved to be a mathematician of exceptional ability and on the second day of the Senate House Examination in 1778 was placed at the head of his year. He graduated as Senior Wrangler and first Smith's prize man and was elected to a fellowship in his college.

The college records of the period are sketchy, but it seems probable that Farish was made a junior tutor soon after his election to the fellowship. He was ordained in December 1780, made Praelector of the college in the following January and admitted to the M.A. degree in the same year. Beyond the fact that he prepared a number of men for ordination little is known about his work as tutor, but it is clear from subsequent events that his special interests lay in the two main fields which characterized the industrial revolution, the substitution of mechanical power for human labour and the application of science to industry. He made a study of the machines in common use, the production of raw materials and all branches of civil engineering. He read widely on these subjects but was not content with second-hand information. During vacations he travelled over the country visiting factories, harbour works and mines and discussing technical matters with anyone willing to talk to him and thus amassing a prodigious store of miscellaneous knowledge. The *University Calendar* for 1802, which was the first to describe the professorial lecture courses in any detail, states that Farish had 'taken an *actual* survey of almost everything curious in the manufactures of the Kingdom'.

He was moderator for several years from about 1790 and proctor in 1792 and 1793, and in both capacities was responsible for the conduct of the exercises and the Senate House Examination. Although the latter had been held for over half a century no satisfactory method of placing the candidates in order of merit had been evolved, and it was

Farish who brought about the obvious reform of awarding marks for individual questions. He was considered one of the most attentive and business-like proctors of his day, and the discipline of the university was never more strictly kept than during his term of office.

In 1792 the Jacksonian Professorship became vacant on the retirement of Isaac Milner. Farish applied for the chair but was defeated in the election by F. J. H. Wollaston, who obtained 35 votes against his 30. However, another opportunity for promotion presented itself two years later when Isaac Pennington resigned from the Professorship of Chemistry to become Regius Professor of Physic, and in January 1794 Farish was elected by open poll of the Senate to succeed him. As lectures on chemistry were already being given by Wollaston, Farish decided to strike out on a line of his own and chose as his subject 'The Application of Natural Philosophy, Natural History, Chemistry, etc. to the Arts, Manufactures and Agriculture of Great Britain'. By arrangement with Wollaston he agreed to lecture on alternate days during the Lent and Easter terms in the Jacksonian lecture-rooms in the south-east corner of the old Botanic Garden where the Zoological Laboratory now stands. The room provided accommodation on benches for 150 people, and gave ample space for the lecturer to display apparatus and diagrams. Smaller rooms adjoining were used as offices and stores, but the whole building was damp, and serious damage sometimes occurred to the equipment and specimens stored there.

Farish began his lectures in the Lent Term of 1795, and in the following year published the 'Plan' of his course in the form of a printed booklet interleaved with blank pages for the use of students. The Plan was reprinted without any substantial alteration in 1803, 1813 and 1821 and was summarized in every *University Calendar* from 1802 to 1837, the year of Farish's death. It consisted of 355 subject headings, but unfortunately there is nothing to indicate the number or content of individual lectures. The course was divided into four parts, and although the field covered was so wide that some parts of it must have been treated rather superficially it may well represent the first serious attempt in any British university to study the industrial implications of a rapidly developing technology. The knowledge required to make such a course successful had been built up during the sixteen years of his tutorship, and some at least of the public lectures which he gave as professor may have taken shape in the college class rooms.

The first part of the printed syllabus was entitled 'Metals and Minerals'. Farish began his course with an account of the earth's structure and the conditions in which the useful metals and minerals are found in nature and then passed to methods of mining and smelting metallic ores and the operation of blast furnaces, giving many examples of the use of the finished products in industry. Lead, silver and copper were treated in a similar fashion, and references to copper-plate engraving, etching, aquatinting and lithography indicate Farish's interest in the arts as well as the manufactures of his day. He then turned to the minerals and described the mining of coal, the quarrying of stone, the production and uses of many non-metallic substances, the manufacture and decoration of pottery and the making of every type of glassware from plate windows to laboratory equipment and optical lenses. An account of the making of gunpowder, acids and alkaline salts brought this wide-ranging survey to a close.

Part II, 'Animal and Vegetable Substances', opened with a section on agriculture, the draining and manuring of land, the breeding of cattle and the growing of corn and timber. Then Farish passed to oils, fats and the whaling industry, and thence to cod, herring and salmon fishing, the tanning of leather, refining of sugar and making of rum. A more detailed description of the manufacture of clothes and fabrics from flax, cotton, silk and wool came next, with diversions on the care of silk-worms and the selective breeding of sheep. The section ended with an account of paper-making and the design of glazing and rolling mills.

Part III, 'On the Construction of Machines', soon came to be considered the most important part of the whole course. It was a natural development from previous courses on experimental philosophy in which the theory and construction of mechanisms had been used to demonstrate the principles of Newtonian mechanics, but it covered a far wider field and went into much greater detail. It opened with a general description of the sources of motive power then available—animal strength, wind, water and steam—and went on to describe the construction of wind and water mills. A more theoretical section on traditional lines dealt with methods of measuring force and velocity. The rest of Part III described the application of these principles in practice. Steam engines, boilers, condensers, air pumps, gauges and valves, piston rods and beam connections, gears and jacks, lathes and cranes were described in detail and illustrated by many working models. There was nothing superficial or slip-shod about this part of the course; Farish was an expert on the machines which were then in use, and it seems fair to claim this as the first real course on mechanical engineering to be given in any British university.

Part IV, 'Water Works and Navigation', dealt mainly with water supplies and the construction of canals, harbours and docks. It also treated of the building and launching of ships, the theories of stability and water resistance, the design of rudders, keels, masts and yards, and ended with a note on tacking and wearing and the handling of ships at sea. There are no grounds for thinking that Farish had any practical experience of seafaring, but all that he taught could be learnt from books and observation, and the long years of naval warfare which followed the outbreak of the French Revolution gave a topical interest to these matters.

Farish not only broke new ground in the substance of his lectures, but also introduced new methods of illustrating them. He provided himself with a large stock of brass gear wheels of various sizes, axles, bars, screws, clamps, etc., and with these built up on the bench in front of his audience working models of the mechanisms he was describing. Among his more permanent pieces of equipment were a steam-engine and a water-wheel with which he set his machines in motion, and models designed by him and constructed under his supervision, of cotton mills, looms, saw-mills and machines for rolling iron and boring cannon, besides numerous drawings and diagrams. His lectures soon became immensely popular and attracted large audiences. They were not intended as a professional training for engineers, but, in the words of the *University Calendar*, 'to excite the attention of persons already acquainted with the principles of mathematics, philosophy and chemistry to REAL PRACTICE; and by drawing their minds to the consideration of the most useful inventions of ingenious men, and in all parts of the Kingdom, to enlarge their sphere of amusement and instruction, and to promote the improvement and progress of the Arts'.

An amusing account of these lectures 'as I remember them some time ago' was published in *Facetiae Cantabrigiensis* by 'Socius' in 1836 over the initials K.Q.M.

Come with me to hear Professor Farish; the hour will be well employed. The experimental philosopher has laid out all his apparatus of cog-wheels, cylinders, bars, pulleys, cranks, screws, blocks, etc., and, with a complacent smile, is contemplating the ingenious combination of all the parts. In the simplest, almost approaching to infantine, manner, he explains all the intricate modes by which these wheels work upon one another, their multipliers, their momentums, and their checks. His sawing-machines, his hat manufactury, his oil press, and cannon foundry, are abundant sources of entertainment. In the latter we see the whole process, from the casting to the firing off of the instrument of war. His explanations of the art of mining and ship-building are perfect in clearness and precision; and the air of simplicity which he throws over the whole is such that the student cannot but smile at the seeming facility of the subject, and the serene indifference with which the Professor treats of the most complex machinery. Under all this appearance of simplicity, it is discoverable that he is a great man. He is one of the best mathematicians the last age produced... Whoever is destined to occupy any situation of distinction, or public utility, cannot do better than add to his stock of information the matter of these very improving lectures; he can never go unimproved away; he will carry with him into life so much ingenious knowledge, if he has given his attention to the course, that he will everywhere meet with consideration and respect, while he can render service or furnish instruction.

In 1798 Farish became President of Magdalene, and two years later vicar of St Giles, the church which faces the college across the Chesterton Road. About this time he married, and some years later moved into Merton House, which he built on a plot of land just south of the junction betwen Queen's Road and Madingley Road. He became renowned for his benevolence and liberal support of local charities. He was treasurer of the Cambridge Charity Schools, and was largely responsible for the building of many new ones. As an acknowledged leader of the evangelical movement in Cambridge he drew large congregations to his church and the old Norman building soon became too small to hold them. He built on several extensions and then, in order to make himself heard in the irregular building which resulted, erected over the three-decker pulpit a parabolic sounding board of his own design which was said to resemble an inverted coal-scuttle. He also indulged his mechanical ingenuity in Merton House by putting in a movable partition wall worked by pulleys and counterweights which could sub-divide either a ground floor room or the one above it. In his old age he was notoriously absent-minded, and on one occasion after bidding goodnight to visitors of opposite sexes who were occupying the two upper bedrooms he started work in the room below on some mathematical calculations. As the fire in the grate died down he decided to make himself more snug and, forgetting his guests, pulled down the partition. Their astonishment on waking up to find themselves in a double-bedded room may well be imagined.

In 1813 Wollaston resigned the Jacksonian Professorship and Farish was elected to succeed him. He made no major alterations in his lectures on changing his chair, though he laid more stress on the increasing use and importance of steam engines. He took over the Jacksonian rooms completely and lectured there on five days a week during the Lent and Easter terms. As the years passed he found it difficult to keep pace with the rapid developments in technology, but he foresaw the time when steam would supply the main motive power for travel by land and sea and the possibility of the discovery of some new agent which would enable man to voyage by air as well. Robert Willis, who was to succeed him, attended and admired his lectures in the middle twenties, and wrote of him in 1851, 'The forms and constructions of manufacturing mechanism underwent so total a change after this course had been arranged, that it is no disparagement to its ingenious author to say that this apparatus, framed in accordance with the methods used in this country in 1813 became useless long before 1837 as a representation of British industry.' However, his lectures retained their popularity and educational value for many years, and continued to attract large audiences. He possessed great gifts as a conversationalist, and his wide interests, vast knowledge and unassuming manners made an evening in his company an experience to remember.

He carried on his university and parochial work until 1836 when he resigned the living of St Giles and retired to a country parsonage at Little Stoneham in Suffolk. He died there in the following year at the age of seventy-seven.

GEORGE BIDDELL AIRY, 1801–1892

From the end of the eighteenth to the middle of the nineteenth century the teaching of engineering in the university was mainly in the hands of the Jacksonian Professors, but valuable contributions were also made by certain holders of the Lucasian Chair of Mathematics and the Plumian Chair of Astronomy. Samuel Vince, who was Plumian Professor from 1796 until his death in 1821, carried out important work in the field of friction, devised a set of experiments to demonstrate phenomena in electricity and magnetism, and illustrated his lectures on mechanics, hydrostatics and optics by the use of various machines and 'philosophical instruments'. He was succeeded by Robert Woodhouse, whose importance in the pre-history of the Engineering Department lies less in his lectures, which followed the established pattern in experimental philosophy, than in his work as leader of the Cambridge Analysts, a society of undergraduates founded in 1812 by his three brilliant disciples, George Peacock, John Herschel and Charles Babbage. Together the four men introduced into England the continental notation for the calculus in place of Newton's fluxions and brought Cambridge back into the main stream of mathematical study. Later Peacock, who became Lowndean Professor and Dean of Ely, played an important part in the Royal Commission of 1850 whose work has already been described. Babbage, Lucasian Professor from 1828 to 1839, is memorable chiefly for his research on calculating machines, which absorbed £17,000 of public money and £6,000 of his private fortune. The great 'analytical engine' with which he struggled for many years was never completed, but parts of the pioneer model were shown in the International Exhibition of 1862 and are now in the South Kensington Science Museum.

Woodhouse was succeeded as Plumian Professor by G. B. Airy, who deserves more than cursory mention as a teacher of engineering because of the part which he played in establishing the study of the theory of structures in Cambridge and in developing closer relations between professional engineers and the university.

George Biddell Airy was born at Alnwick in Northumberland on 27 July 1801. His father was a Collector of Customs, but he failed to make a success of his profession and after being moved several times lost his post in 1813 and was reduced to a state of comparative poverty. George's mother, however, was a woman of strong character, and with the help of her brother, Arthur Biddell, a prosperous farmer and valuer who lived at Playford near Ipswich, was able to provide her three children with a good education. After attending elementary schools in Hereford George was sent to Colchester Grammar School, where he distinguished himself by his high intelligence and retentive memory. He claimed that at the age of fifteen he once repeated 2,394 lines of Latin and Greek poetry, 'probably without missing a word'.

In 1819 he was admitted as a sizar to Trinity College, where he soon came to be recognized as the outstanding scholar of his year. Before the end of his first Michaelmas Term he was awarded a £20 exhibition and in the following May was appointed Head Lecturer's Sizar with an annual stipend of £10. In the college examinations at the end of his first and second years in residence he was easily first and in April 1822 was awarded a scholarship on the foundation. He had been coaching private pupils for some years and was now, to his great satisfaction, financially independent. His first 'act', which he kept on 12 February 1822, attracted great attention, and it came as no surprise to the university when, in January 1823, he graduated as Senior Wrangler and first Smith's prize man. In the following year his college elected him to a fellowship and appointed him Assistant Mathematical Tutor. His duties included lectures to undergraduates in their second and third years and a special course of exercises for those who were to take their degrees in the following January. He examined for the college in mathematics, *Newton* and optics. He also supervised the work of three or four private pupils and did a great deal of private research in the fields of mathematics and optics. In the vacations he travelled extensively, sometimes with his sister, sometimes on study tours with his pupils, and wherever he went he missed no opportunity of studying engineering works—particularly bridges, which had become, as Milner might have said, his 'hobby-horses'.

His interest in engineering began in boyhood and was due largely to the influence of his uncle, Arthur Biddell, who took a great interest in George's education and welcomed him at his home at Playford during the school holidays. Biddell had a good library from which the boy derived an enormous amount of miscellaneous information, but in his search for knowledge he did not confine himself to books. At an early age he made an accurate drawing of a threshing machine, and when a six horse-power steam engine at a neighbouring brewery was opened for inspection he had the 'extreme felicity' of examining its working parts. Experiments with the object glass of a pair of opera glasses aroused his interest in optics and by the time he entered the university he claimed to know more about the action of lenses in a telescope than most opticians. It was always his practice to record in writing matters of interest which he encountered in his reading or on his travels, and as an undergraduate he kept notes on the construction of steam engines, the design of bridges and many other mechanical matters. Soon after graduation he was elected to the Cambridge Philosophical Society and in 1826 he read a paper there on the steam engines and pumps in use in Cornish mines and another on the theory of pendulums, balances and escapements. In the same year he presented a paper on the ellipticity of the earth to the Royal Society.

He was now a major fellow of Trinity and seemed to be committed to an academic career, but he was determined neither to take Holy Orders nor to remain unmarried. While on a walking tour in Derbyshire he had fallen in love with Richarda Smith, eldest daughter of the chaplain to the Duke of Devonshire, and he intended to ask for her hand in marriage as soon as he could provide a home and income worthy of her. Until he could do so her father refused to sanction any formal engagement.

In October 1826 the Lucasian Chair of Mathematics became vacant and after some hesitation due to the low stipend—only £100 a year—he applied for it and was unanimously elected. Two years later he was able to transfer to the Plumian Professorship, which carried a stipend of £300 and a house in the Observatory. However, he did not feel that even this would meet his needs if he were to marry Richarda and he petitioned the Vice-Chancellor for an increase, stating his case in a letter which he circulated to every resident member of the Senate. After an inquiry which lasted several months the University agreed to increase the stipend to £500 by an annual contribution from the Chest, and as soon as Airy received his first payment he set off for the Derbyshire rectory. His formal proposal of marriage was accepted and the ceremony took place on 24 March 1830. The marriage was a very happy one. Richarda was considered a great beauty, and she was also very intelligent, well-educated and of a most friendly and hospitable disposition. Nine children were born to them between 1831 and 1845.

As Lucasian Professor Airy revived the lectures on Experimental Philosophy. They had not been given for many years and there were many difficulties to be overcome. No lecture-room was appropriated to the Professorship and the apparatus with which experiments had formerly been conducted had long since disappeared. However, he eventually obtained a room under the University Library which had been used by Vince and there, in the Lent Term 1827, he delivered his first course of lectures on bridges, trusses, mechanics and optics. Sixty-four men attended the course, but their fees were barely sufficient to cover expenses and after two years he complained that his total income from the lectures had only exceeded his expenditure on them by £45.

Although as Plumian Professor his duties were mainly astronomical he continued to deliver his course on experimental philosophy, which evidently met a recognized requirement. The syllabus was printed in 1832. The headings were Mechanics (subdivided into Statics and Dynamics), Hydrostatics and Pneumatics, Optics, Undulations and Polarization. Under the heading Statics he dealt with toothed wheels, wedges, screws, pulleys, etc., and then passed on to his main engineering interest, which was in structures. As examples on the theory of roofs he chose the Sheldonian Theatre at Oxford, the Cambridge Observatory, the Hall of Trinity College, Cambridge, and Westminster Hall. To illustrate the theory of the arch he described Blackfriars, Waterloo and London bridges, and for the theory of domes the cathedral at Florence, St Peter's in Rome and St Paul's in London. On groins he used the fan vaulting of King's College chapel as an example, and for suspension bridges that over the Menai Strait. In the lectures on Optics he made use of many instruments of his own design, and diagrams of interference phenomena were drawn for him by his wife. His lectures were crowded into four weeks in the easter Term, and on average were attended by about fifty students. Besides having a general educational value the course aimed specifically at the questions asked in the Senate House examination.

Airy's work at the Cambridge Observatory soon attracted world-wide attention, and in October 1834 he was offered and accepted the office of Astronomer Royal. He resigned from the Plumian Professorship and moved to Greenwich in the following year, but for the rest of his long life he continued to keep in close touch with the affairs of the university. He also acquired a great reputation as a consulting engineer, and his son Wilfred, in an introductory note to his father's autobiography, wrote

> Any subject which had a distinctly practical object and could be advanced by mathematical investigation, possessed interest for him...He was extremely well versed in mechanics, and in the principles and theory of construction, and took the greatest interest in large engineering works. This led to much communication with Stephenson, Brunel and other engineers, who consulted him freely on the subject of great works on which they were engaged: in particular he rendered much assistance in connection with the construction of the Britannia Bridge over the Menai Straits.

He was elected a Fellow of the Royal Society in 1836, won the Society's Gold Medal ten years later and was President in 1872–3. He became a member of the Institution of Civil Engineers and was awarded the Telford Medal in 1867 for a paper 'On the use of the Suspension Bridge with Stiffened Roadway for Railway and other Bridges of Great Span.' After three times refusing a knighthood he became a Knight Commander of the Bath in 1872.

James Stuart, the first Professor of Mechanism at Cambridge, knew him well during the last twenty-five years of his life, and a character sketch from Stuart's 'Reminiscences' is worth quoting.

> Sir George Airy was one of the most remarkable men, indeed I think I may say *the* most remarkable man, whom I ever met. In every part of his thoughts he was the most absolutely original person with whom I ever came in contact...the whole of the Observatory was full of his inventions—doors which shut by contrivances of his own, arrangements for holding papers, for making clocks go simultaneously, for regulating pendulums, for arranging garden beds, for keeping planks from twisting, for every conceivable thing from the greatest to the smallest. On all there was the impress of an original and versatile mind, bubbling over with inventiveness. His conversation ranged over everything of

human interest, historical, religious or scientific, and his ideas had the immense merit of being always couched in the most simple and intelligible language. He was a colossal-minded man, and his ideas seemed to be executed in granite.

Airy held the office of Astronomer Royal for forty-six years. He retired in 1881 and died at Greenwich on 2 January 1892 at the great age of ninety-one. He was buried in Playford churchyard.

ROBERT WILLIS, 1800–1875

The Rev. Robert Willis, Jacksonian Professor from 1837 until his death in 1875, occupies a unique position as a teacher of engineering before the establishment of the chair of Mechanism. He was the first to give lecture courses intended specifically for candidates for a degree examination in applied science and the first Cambridge professor to win an international reputation as a mechanical engineer.

He was born in London on 27 February 1800. His father, Dr Robert Darling Willis, and his grandfather, the Rev. Francis Willis, were pioneers in the humane treatment of lunacy and both attended George III during his periods of insanity. Robert Darling was a junior Fellow of Gonville and Caius College until his death in 1821 and as such was not allowed to marry, but he had two children, Robert and Mary, both of whom took his surname. Nothing is known about their mother or the environment in which they were brought up. Mary married the Rev. William Clark, Professor of Anatomy at Cambridge, and their son, John Willis Clark, of whom more will be said later, was for many years Registrary of the University. Robert was a sickly child and was educated at home until he reached the age of twenty-one, when he was sent to King's Lynn to work for a year under a clergyman named Kidd.[1]

Robert Willis was a talented musician and a good draughtsman, and from an early age his chief hobby was the study of ancient buildings. When only eighteen he took out a patent for the improvement of the pedal harp, and soon afterwards wrote a paper which attracted some attention by exposing a fraudulent 'automaton chess player' which in fact had a man concealed inside it. In 1822 he was entered as a pensioner in Caius College, his father having died about a year earlier.

[1] Probably Thomas Kidd of Trinity College, who graduated in 1794.

He graduated as ninth wrangler in 1826 and was at once elected to a fellowship. He was ordained in 1827, as he wished to make his career in the university, but he never held a church living and four years before his death, when the religious tests were abolished, he relinquished Holy Orders by deed poll.

As a fellow of Caius he devoted himself primarily to the study of subjects in which physics and animal mechanisms could be treated mathematically. In 1829 his paper on vowel sounds and reed organ pipes was published in the *Transactions of the Cambridge Philosophical Society* and in the following year, at the age of thirty, he was elected Fellow of the Royal Society. In 1831 he joined the British Association for the Advancement of Science, which was just beginning its career under the presidency of Earl Fitzwilliam, and wrote a report on the phenomena of sound which was read at a meeting at Oxford. In 1832 he married, and on returning from his honeymoon on the continent wrote a paper entitled 'The Architecture of the Middle Ages, especially in Italy.' It was the first of many important publications on architecture and archaeology.

In 1837 Farish died and Willis was elected unopposed to succeed him in the Jacksonian Chair. As an undergraduate he had attended Farish's lectures and his personal copy of the printed plan of the course, with notes and sketches on the interleaved pages, is preserved in the University Library. He adopted Farish's method of illustrating his lectures by means of models built up of component parts, but confined the scope of his course to the scientific study of mechanism. He designed and constructed an entirely new range of models, including sets of gear wheels cut to the same generating circle and the same pitch so that any two wheels would intermesh. He pointed out the practical advantages of this method in a paper which he read to the British Association in the year of his election to the professorship. In the following year he developed the study more fully in a paper read to the Institution of Civil Engineers and explained the construction of the 'odontograph', a device for obtaining the centres from which two intermeshing gear wheels could be cut. He constantly modified his models in the light of experience, and in 1851 published a full description of them in a book entitled *A System of Apparatus for the use of Lecturers and Experi-*

menters in Mechanical Philosophy which embodied the results of fourteen years of teaching. It gave details of the dimensions and scantlings of every part so that the models could be exactly copied, and the text was illustrated by plates in which figures of the models and their component parts were beautifully drawn to scale in isometric projection.

The printed syllabus of Willis's lectures was first published in 1841 under the title 'A Course of Experimental Lectures on the Principles of Mechanism.' It was divided into two parts, which dealt respectively with 'Trains of Mechanism' and 'Mechanism considered with respect to Force.' In it Willis developed a scientific classification of machines from a purely kinematic point of view without reference to the forces at work or the energy transmitted. His theories were further developed and elaborated in a textbook called *Principles of Mechanism* which was published in the same year. It was immediately recognized as the most complete treatise which had yet appeared on the science of mechanism, and placed him in the forefront of the teachers of this subject. In the preface Willis argued that although the science of mechanism had occupied the attention of many eminent mathematicians no one had ever attempted to discover

> the principles upon which is to be based a science that will enable us either to reduce the movements and actions of a complex engine to system, or to give answers to the questions that naturally arise upon considering such engines;—for example, are the means by which the results are obtained the best that might have been employed? Or what are the various methods that might have been substituted for them? Yet there appears no reason why the construction of a machine for a given purpose should not, like any usual problem, be so reduced to the dominion of the mathematician, as to enable him to obtain, by direct and certain methods, all the forms and arrangements that are applicable to the desired purpose, from which he may select at pleasure. At present, questions of this kind can only be solved by that species of intuition which long familiarity with a subject usually confers upon experienced persons, but which they are totally unable to communicate to others.

Except for a brief period during the 1840s Willis lectured only during the Michaelmas Term. His course was described in the *University Calendar* as 'On Statics, Dynamics and Mechanism, with their practical applications to Manufacturing processes, to Engineering and Architecture.' The lectures became immensely popular, and for many years drew crowded audiences. His nephew John Willis Clark wrote that

as a lecturer he had extraordinary gifts. He used neither manuscript or notes, but whether he was describing a machine or a building an uninterrupted stream of lucid exposition flowed from his lips, carrying his hearers without weariness through the most intricate details and making them grasp the most complex history of construction...No matter how dry the subject, he knew how to make it interesting, and whether he discoursed on rope-making or the organ, on joints or the Jacquard loom he held his audience spell-bound, and dismissed them charmed alike with the knowledge they had gained, and the pure English in which it had been conveyed to them.

This was the hey-day of professorial lecturing at Cambridge when a number of outstanding teachers expounded the exciting discoveries which were being made in all branches of science. The subject of their courses often had little to do with the work required for admission to the B.A. degree, but men who did not aspire to a high place in the list of wranglers could well afford to devote some of their time to matters of general educational interest. The university had doubled in numbers since the turn of the century and a new spirit of interest and inquiry was abroad. Men flocked to these lectures because they offered information about important topical subjects outside the narrow syllabus of the Senate House Examinations.

Unfortunately for the professors this happy state of affairs did not last for many years. As described in the first chapter the monopoly of Newtonian mathematics as the main requirement for the B.A. degree was broken and the modern system of triposes gradually replaced the single Senate House Examination. The standard required for admission to the ordinary degree was also raised, and although this was a slow process its effect was making itself felt by the middle forties. A letter published in *The Times* in 1846 gave a depressing account of the scanty attendance at professors' lectures at Cambridge, and made special reference to Willis, whose course had formerly been so well attended.

The Professor of Experimental Philosophy is in the worst plight of all. At any other place the attraction of machinery is irresistible. The engine room of a Margate steamer is choked up with enquiring heads, and an acting steam coffee-mill in the window will treble the customers of any grocer in Whitechapel. But

Cambridge birds are not to be so caught. Low-pressure steam engines and long-action organs are exhibited in vain. The official income of years has been sunk in the apparatus, but the academical public disregard the tremendous hit, and the powers of mechanism are dolefully developed to four men and a boy.

This was probably an exaggeration even when it was written, but in any case matters soon began to improve. When Whewell's scheme for relating the professorial lectures more closely to the undergraduate examinations came into effect in 1849 Willis began a new course on mechanics designed especially for students reading for honours in mathematics. Later he gave a course of six lectures a week during the Michaelmas Term for the poll-men on 'Mechanics and Mechanism and their application to Manufacturing Processes, the Steam Engine, etc.', in preparation for the 'Professors' Examination' in February.

The events which led to the appointment of a Royal Commission in 1850 have already been described. Willis took no active part in the bitter controversies which divided the university at this time, but continued quietly with his work. His lectures on Mechanism received high praise from the commissioners, who wanted to see civil engineering accepted as one of the new triposes and a Professorship of Engineering established. A group which favoured this idea, led by Henry Latham of Trinity Hall, wished to put Willis into the new chair, but the scheme had to be abandoned for lack of funds. The surviving records of this critical time are unfortunately incomplete, but it seems certain that Willis showed no desire to give up the comparative freedom of the Jacksonian Chair in order to undertake the thankless task of building up a school of technology for which neither the university nor the profession of engineering was yet ready. His time and energy were fully occupied in other directions.

His reputation outside Cambridge as an authority on the science of mechanism and its application to problems of civil engineering was growing rapidly. Membership of the Royal Society and the British Association brought him into contact with many of the country's foremost scientists and engineers. Thomas Telford, the great builder of roads, bridges and canals, had been President of the Institution of Civil Engineers since 1820 and a Fellow of the Royal Society since 1827. He used his immense prestige to establish a close liaison between the old-established Society and the virile young organization of practising engineers, and it was doubtless with his approval that Willis was elected an Honorary Member of the Institution in May 1838.

In 1849 Willis was appointed a member of a Royal Commission to inquire into the application of iron to railway structures. Experiments were carried out in Portsmouth Dockyard to determine whether the deflections of bridges were affected by the velocity of loads moving along them. Willis was largely responsible for an appendix to the main report in which some of the conclusions reached by experiment were confirmed mathematically, and to demonstrate the phenomena under investigation he designed and constructed an apparatus which was subsequently exhibited at a meeting of the British Association. In 1851 he was chosen as one of the jurors in the Great Exhibition in Hyde Park and was responsible for the report on manufacturing machines and tools. Later he gave a lecture in a course on the exhibition organized by the Society of Arts, in which he advocated a proper system of professional education for engineers and pleaded for 'a more intimate union and greater confidence between scientific and practical men by teaching them their wants and requirements, their methods and powers, so that peculiar properties and advantages of each may be made to assist in the perfection of the other'. To further this object he accepted the post of lecturer on applied mechanics when the government School of Mines was established in Jermyn Street, London, in 1853. For many years he delivered a course of thirty-six lectures there annually, with a supplementary course of six lectures specially designed for working men which he gave in alternate years. In 1855 he was appointed Vice-President of the Paris Exhibition and reporter for the class of machinery for making textile fabrics—a proof of the reputation he had gained from his work for the exhibition in London. For this service he was awarded the Cross of the Legion of Honour by Napoleon III.

Together with all this activity in the field of engineering he carried out a great deal of original research into architectural history and the methods of construction of ancient buildings. He became an authority on medieval Latin and the handwriting of scribes, and published many papers on the English cathedrals and other buildings for which he was awarded the Royal Gold Medal of the Institution of Architects. In 1861 he delivered the Rede Lecture on the 'Architectural History of the University of Cambridge' to a crowded audience in the Senate House. His work in yet another field, that of astronomy, led to his nomination by the President of the Royal Society as Visitor to the Royal Observatory at Greenwich in succession to William Whewell. He was President of the British Association in 1862 when it met in Cambridge, and in the following year, in Newcastle, he presided over its mechanical section.

In 1865 the regulations for the ordinary degree were completely recast. The poll-men were required to take a 'General Examination' and one of five 'Special Examinations' which superseded the Professor's Examinations. One of the 'Specials' was on Mechanism and Applied Science, and the schedule for it covered the principles of mechanism, the steam engine, horology, heat, electricity and magnetism. Lectures on the last two subjects were given by the Cavendish Professor of Physics and the remainder by Willis. In the same year Willis was allocated a suite of rooms in the new building which had been erected in the Botanic Garden. He found his new quarters spacious and convenient, but his collection of models, which had suffered much injury from the dampness of the old Jacksonian lecture-room, was further damaged by its hurried removal to the new museum before its walls were dry. In time, however, the models were repaired and set up in glass cases.

In 1870 Willis published a revised and expanded edition of his textbook on the principles of mechanism, which was by this time a standard work for all university students of engineering. Two years later failing health compelled him to ask for a deputy to carry out his professorial duties and the Senate approved the appointment of the Rev. J. C. Williams Ellis, of Sidney Sussex College, who delivered the Jacksonian lectures for the next two years and received one-third of the stipend. In February 1875 Willis was attacked by bronchitis, and on the last day of that month he died. A memoir published in the minutes of proceedings of the Institution of Civil Engineers described him as 'one of the last of the brilliant circle of Cambridge professors who, from the varied nature of their acquirements, were the glory of their University from the middle period of the century'.

This is a fitting tribute to his versatility, but it is as a scientific engineer that he should be chiefly remembered. His academic research and teaching derived directly from Newton and the Industrial Revolution and were entirely in the Cambridge tradition. He built on foundations which had been laid by Milner, Farish, Airy and their predecessors. He was no innovator, but a great consolidator, and when he died the formal recognition of engineering as a proper subject for study, research and undergraduate teaching was accepted almost without question by a large majority in the university.

3 JAMES STUART

Professor of Mechanism and Applied Mechanics, 1875–1890

EARLY YEARS, 1843–1875

James Stuart was born on 2 January 1843 at Balgonie in Fifeshire, where his father, Joseph Gordon Stuart, owned and managed a prosperous linen mill. Joseph Stuart, the son of an Edinburgh lawyer, had studied at Aberdeen and Edinburgh Universities. He was an active politician with advanced liberal views and unlike many of his contemporary employers took a keen and practical interest in the welfare of his work people. James' mother, a deeply religious woman whose stern Calvinistic faith mellowed as she grew older, exerted a strong influence on her children and after the death of her husband became James's most intimate friend and companion.

When James was eight years old he went to live with his maternal grandmother at St Andrews, where he attended the Madras College and obtained a good grounding in English literature, Latin and the elements of engineering drawing. He returned to Balgonie after three years and was put under the charge of a private tutor who taught him Latin, Greek and French but concentrated mainly on mathematics, in which he showed exceptional ability. He spent part of each day in his father's counting house learning office work and book-keeping and in the workshops attached to the mill, where he was put through a kind of unofficial apprenticeship. When twelve years old he started a drawing class for the workshop apprentices in the family kitchen—an early venture into the education of young mechanics which was to be one of his main interests in later life. In 1859, when he was nearly seventeen years old, his father took a house in St Andrews and James began work in the university. On account of the admirable grounding he had received from his tutor he was allowed by the Senatus to dispense with the first year's course and at the end of his second year he was admitted to the B.A. degree. He won a Ferguson Scholarship in Classics and

Mathematics, but instead of taking this up at a Scottish university for further study he entered for a minor scholarship at Trinity College, Cambridge. He was again successful, and matriculated there in the Michaelmas Term 1862.

As an undergraduate he worked and played hard. He coached with the famous E. J. Routh of Peterhouse and stroked the first boat of the Second Trinity Club, of which he was treasurer. He was elected to a foundation scholarship in 1864 and two years later graduated as third wrangler and was awarded a college fellowship. His father died towards the end of that year and James spent some months at home putting the family affairs in order. He returned to Cambridge in February 1867 and took up the post of Assistant Tutor in Trinity, but he continued to play an active part in the management of the Balgonie mill until it passed out of the hands of the family in 1884.

Shortly after taking his degree Stuart conceived two ideas which provided the principal driving force behind his main activities until he was elected to the chair of Mechanism. One related to the internal teaching methods of the university and the other to the extension of its influence outside Cambridge. By this time the new examinations for both honours and poll-men had been established, and the smaller colleges were finding it impossible to provide the teaching needed to cover the extended syllabuses. Stuart played a leading part in organizing a system of inter-collegiate lecture courses very much on the lines suggested by the Royal Commissioners of 1850. He himself gave the first course in 1868, taking as his subjects physical astronomy and the mathematical theories of electricity, magnetism and heat. His lectures were well attended and after some years the practice became general and was officially recognized by both the colleges and the university.

Stuart's other innovation was the institution of the extra-mural courses for which Cambridge is now famous. He had told his mother in 1866 that it was his ambition to found 'a sort of peripatetic university' whose professors would circulate among the big towns and carry the torch of academic learning into the main industrial regions. His opportunity came in the summer of the following year, when he was invited by the North of England Council for promoting the Higher Education of Women to lecture to classes of ladies in Manchester,

Liverpool, Sheffield and Leeds. The President of the Council was Mrs Josephine Butler, one of the chief protagonists of the new movement for securing women's rights, and the secretary was Miss A. J. Clough, who later became the first Principal of Newnham College. These ladies became Stuart's intimate friends, and from this time he was one of the champions for women in their struggle for emancipation.

His lectures to the ladies were followed by one to the railway workmen at Crewe, which he gave at the request of W. M. Moorsom, an old Trinity friend on the technical staff of the railway who had recently helped to found a Mechanics' Institute in that district. No less than fifteen hundred workmen assembled to hear him, and his reception was so enthusiastic that he agreed to go back in the following summer and give a course of weekly lectures spread over a period of two months. From that and similar engagements he soon aroused great interest in his scheme, and by 1871 he felt that sufficient progress had been made to justify an appeal to the university for formal recognition. At his suggestion memorials were sent to the Vice-Chancellor and the Council of the Senate by the North of England Council, the Crewe Mechanics' Institute, the Rochdale Equitable Pioneers' Society and the mayor and other inhabitants of Leeds, all requesting the university to institute a regular system of lectures in the provinces. Stuart accompanied these memorials with a letter addressed to all resident members of the Senate urging the appointment of a syndicate to consider how assistance might be given in furthering the higher education of those classes in the great towns who were debarred from residence in a university. The Senate accepted the proposals and appointed a syndicate, with Stuart as secretary, which reported in 1873 that there was enough evidence of the wish for higher education in co-operation with the university to justify an experiment being made. In October of that year the first courses under university auspices were given in Nottingham, Derby and Leicester, and they were so successful that before the experimental period had elapsed the syndicate reported in favour of making the scheme permanent. The Senate agreed, and responsibility for the future conduct of the lectures was given to the Local Examinations Syndicate, one of the permanent executive committees of the university.

Stuart remained honorary secretary and continued to lecture in the provinces as well as in Cambridge. He also continued to direct the management of the Balgonie mill and took an increasing interest in all types of industrial machinery. He made many friends in the engineering industry and was often consulted about technical problems involving the application of mathematics or a knowledge of theoretical physics.

The success of the inter-collegiate and extra-mural lectures made Stuart a marked man in the university, and when the Jacksonian Professorship fell vacant on the death of Willis in 1875 many considered him to be the obvious successor. He was at first disinclined to enter the lists, being aware that there was considerable difference of opinion in the Senate as to the subject which the new professor should teach, but he allowed himself to be persuaded and on 19 March formally offered himself as a candidate to the members of the Electoral Roll. He was careful not to tie himself to any particular subject, and he told his mother that he would rather remain Assistant Tutor of Trinity, where he could choose his own subjects for teaching, than accept a professorship which would confine him to a single branch of natural philosophy. However, he felt unable to stand up against the pressure from his friends.

Whether I be elected or not the gratifying point is that this is the *first* professorship which has become vacant in the subject of my studies since I took my degree, and that so many of the notable people in the University have wished me to get it. For, when I was unwilling to stand for it those who pressed me to do so are the Master of Trinity, the Master of St John's, the Vice-Chancellor, the two divinity professors, besides *all* the tutors and fellows of Trinity College.

In his letter of application Stuart made no mention of the subject which he wished to teach if he were elected, but the two testimonials which he submitted spoke only of his qualifications as an engineer. His friend Moorsom of Crewe wrote:

He and I have passed many hours together in the works here and have on several occasions discussed questions arising out of the mechanical operations that are carried out here; I therefore speak with my own knowledge when I say that he is familiar with machinery in detail and is quite at home in these works... I believe that Stuarts' great success in rousing the interest of our workmen in his lectures was due in large measure to the fact that his mind has a strong practical bent and so his thoughts on mechanical questions are very easily followed by our engine builders.

His second sponsor was R. F. Martin, another contemporary of Stuart's at Cambridge who had become a professional engineer. In his testimonial to the Vice-Chancellor he made a strong plea for the establishment of an engineering school in the university. His own education at Cambridge had been immensely valuable to him professionally as well as otherwise, but the absence of any 'distinctly mechanical' work was a great disadvantage and lost the university much influence.

The demand for this teaching is now enormous...and Stuart is quite the best man I know of, of my own time, at any rate, to undertake it...He has a very considerable knowledge of practical mechanics. His early life has, I know, been spent in a mechanical neighbourhood and early familiarity has much to do with aptitude. I have often consulted him upon engineering matters and have always found he showed a great facility in handling such subjects. Then I think that his last few years, spent so much among mechanical scenes and men, not to add his youth and enormous enthusiasm, are all in his favour.

Stuart left Cambridge for a short time after submitting his application, but finding on his return that feeling had hardened in favour of a chemist as the new professor he immediately withdrew. He explained his motives in a letter to his mother.

I have withdrawn my name from being a candidate for the Jacksonian Professorship. I came here on Saturday and I found that there had been a growing desire to have this chair filled by a chemist...The Jacksonian professorship now vacant is, by the founder's will, more particularly one of chemistry—the practice however of late professors has been to make it one of what may be called Mechanical Science. The desire seems now pretty general to revert to Chemistry and to make a *new* professorship of Mechanical Science; and it seems to me to be the best way, to retire from this professorship and wait for the other, because it is on the mechanical and not on the chemistry ground that people desire me to be a professor. Again there is this point to consider. If I stand and am elected for this present professorship I will teach Mechanical Sciences and not chemistry and there seems little chance that a new professor of chemistry would be made; yet if a chemist is appointed to the present chair there seems little doubt that a professorship of Mechanical Science will be founded. This seems more likely to be the case if I personally am prepared to take it should it be formed, and therefore I think I have acted rightly in retiring at present. As to the phrase 'Mechanical Science' it covers Electricity and such like subjects with which I am acquainted more particularly.

Stuart's forecast was correct, and a syndicate was appointed on 29 April to consider the establishment of a Professorship of Mechanism and Engineering. The post was approved under an amended title and Stuart immediately offered himself as a candidate. In his letter to the Vice-Chancellor he outlined his qualifications.

May I state that for a considerable time both before and after I came to the University I was occupied in the practical superintendence and management of machinery, that for the last seven years I have given lectures in the University on various branches of applied mathematics, and that during the same period I have had frequent opportunities of studying the applications of machinery in large works and have been from time to time in communication with practical enquiries in regard to the solution of technical difficulties.

Voting took place on 17 November 1875 and Stuart polled 111 votes against 81 cast for Routh and 36 for Williams Ellis. The bells of Great St Mary's church rang out to announce his election, as was then the custom, and a large number of his friends went up to his room and led him to the Senate House to be installed.

So I went there in a great hurry and fluster and the Senate House was full of people and I had to shake above fifty people's hands as I pushed on to the Vice-Chancellor's chair where he read a Latin piece out of a book about swearing to do my best for the University and then I said 'Do ita fidem' and had to kneel down when the Vice-Chancellor said 'in name of the authority committed to me I admit you as Professor of Mechanical Science'—this however he said in Latin I kneeling with my hands between his and then the ceremony was complete...[1]

THE YEARS OF EXPANSION, 1875–1884

In embarking on his duties as Professor of Mechanism Stuart had the backing of many influential people in the university, including Dr S. G. Phear of Emmanuel, who had been Vice-Chancellor since October 1874. Phear had graduated as fourth wrangler in 1852 and was now one of the most influential and respected leaders of the reforming party in the Senate. The university was in the middle of a period of rapid expansion which lasted from 1860 to 1890, and he was one of the first to realize what the consequences of this would be. In his address at the end of his first year in office he predicted that as more students would in future

[1] Letter to his mother, dated 20 November 1875.

come from less wealthy homes than formerly the proportion of undergraduates 'unprompted to active study' would decrease and the social influences which encourage industry and simplicity of life would be strengthened. He welcomed the establishment of the chair of Mechanism because it promised to open up a field of instruction likely to attract a class hitherto untouched by university teaching. His views were thus in harmony with those of Stuart when he established the extra-mural courses and with his ambition to provide in his new department an education which would be welcomed by his friends in the industrial areas.

The only assets which Stuart inherited when he took up his appointment as professor were a half share in the Jacksonian lecture-room in the New Museums Building, the use of the two small rooms behind it and the remains of Willis' models, which by that time were in a shocking state of disrepair. He at once set about the task of restoring them, and early in 1876 brought one of the Balgonie craftsmen, James Lister, to Cambridge to act as his assistant. He set up a small smithy at the back of the lecture room and equipped it with Willis's anvil, portable fire and bellows and a small lathe. One of the first items to be repaired was a Jacquard loom which, with a model of an ancient loom made by Lister, he used in his inaugural lecture on 23 October 1876. His original idea was to centre his teaching round a collection of models and small machines, and in his first annual report to the Museums and Lecture Rooms Syndicate in May 1876 he asked for permission to set up a Museum of Mechanical Contrivances on similar lines to the existing Museums of Geology and Comparative Anatomy. 'Such a Museum should contain models of the larger and well-selected specimens of the smaller sized mechanical contrivances, as well as specimens of the various products of mechanical industry, and models and drawings of the various architectural and structural designs.' In a letter to his mother he said he believed that his work as a professor would consist more in providing a groundwork for study than in actual teaching, and he expected to do considerably less lecturing in the future than he had done in the past as a college tutor.

However, a good deal of lecturing had to be done, particularly to candidates for the Special Examination in Mechanism and Applied

HAIGH & HOCHLAND LTD

UNIVERSITY FOREIGN AND
INDUSTRIAL BOOKSELLERS

365 OXFORD ROAD
MANCHESTER 13

Telephone : ARDwick 4156/7

Also at **11a Whitworth Street, Manchester 1** **Telephone : CENtral 9391**

Dr. E. H. Robinson, 13·4·1967

Dept. Econ. Hist.

		@	£	s.	d.
	Hilken/Engineering		2	5	-
	RECEIVED BY				

C 02449

LAMSON PARAGON

Science. The theory of light, heat, magnetism and electricity, on which questions were asked in the 'Special', was now taught in the Department of Physics, and Stuart arranged with Clerk Maxwell, the Cavendish Professor, to take over all the lectures on these subjects. He also transferred to the Cavendish Laboratory all Willis's apparatus connected with acoustics, electricity and light. Stuart himself, after his inaugural lecture, gave a course on mechanism and applied mechanics which carried on into the Lent Term 1877, when he also dealt with the dynamics of machinery, prime movers, governors and the applications of friction. In the Easter Term he lectured on the theory of structures and strength of materials and gave a course on the elements of graphical statics. All these courses were intended as a preparation for the 'Special' in applied science, which had been revised early in 1876. The new schedule for the section headed 'Mechanism and Applied Mechanics' was as follows: 'General principles of Machines and Mechanism or Kinematics. The elements of Applied Mechanics, including the laws and practical application of Friction and Elasticity and Strength of Materials. The elements of Graphical Statics as applied to simple frames. Steam and Steam Engines.' The first examination under the new regulations was held in the Michaelmas Term 1876, before Stuart had started his professorial lectures. Only two men passed, both being placed in the second class, but the number of candidates is not stated.

Stuart's second annual report, published in April 1877, described how, with the help of Lister, he had repaired Willis's models and made a considerable number of new ones. Even so the collection was far from complete, but this did not matter greatly as there was no building in which they could be properly displayed. He repeated his request for a museum and then went on to make a new proposal which was to have far-reaching consequences.

It would be exceedingly valuable for the Engineering pupils who attend my lectures, if there were also a workshop in which they could learn the use of the tools, and a room where they could learn mechanical drawing. I should then be able not only to endeavour to make my pupils acquainted with the present position and history of Mechanical Science, the problems still requiring solution, and the methods of attacking these problems, but to hope also that they might obtain, while at the University, such a knowledge of the first rudiments of practical work as might enable them to shorten their future period of apprenticeship without disadvantage, and thus render it easier for them to come to the University.

Stuart never got a building for his museum, but in the Long Vacation of 1877 he moved all the models into one of the two small rooms behind the lecture-room and fitted up the other as an instructional workshop for four pupils. The course in the Michaelmas Term was so successful that during the Christmas vacation he managed to squeeze three more places into the workshop. Of the seven pupils who worked there during the Lent Term five had already obtained honours in the Mathematical Tripos and two were reading for the degree. Four of the class intended to take up engineering as a profession. To defray the cost of equipping the workshop and pay the wages of two skilled workmen who acted as instructors he began to manufacture pieces of equipment for other departments or members of the university staff. Thus from its inception the Mechanical Workshop combined teaching and profit making.

It was the business of the Museums and Lecture Rooms Syndicate to provide accommodation for the scientific departments, and in the spring of 1878 Stuart persuaded them to ask the university to give him a wooden hut to serve as a workshop and smithy for twenty-five students. If this were done he promised to furnish and equip it at his own expense and run it without cost to the university. On 6 June approval was given to build a hut fifty feet long and twenty wide in the garden behind the Botanic Museum. Work started in July and was completed before the end of the Long Vacation. Stuart bought tools and equipment in Manchester and London and a screw-cutting lathe was presented by an old Trinity friend named Van Sittart. Soon six small foot lathes stood in a row in the new building, carpenters' benches were built round the walls and tools were provided for working in wood and iron. Stuart told his mother that the workshop was a gem, and that his chief workman was in ecstasies about it.

On 7 October 1878 the *University Reporter* announced that the NEW WORKSHOP would be opened for the reception of pupils on Friday 18 October. The notice attracted considerable attention and was copied by many of the newspapers, including some French ones. The retiring

Vice-Chancellor referred in his address to the workshop and smithy 'now, by the munificence of the Professor of Mechanism, abundantly furnished at his own cost with useful valuable machinery and occupied by a large class of eager and industrious students'.

Stuart's private account book[1] shows that the total cost of the workshop up to the opening day amounted to £1,144. 19*s.* 0*d.* and that a further £91. 11*s.* 0*d.* was spent on tools etc. during the Michaelmas Term. These figures include £140 for the value of the site and £303. 2*s.* 3*d.* for the cost of the building and the architects' commission. Apart from these two sums, which were charged to the university, the whole expense was borne by Stuart.

The workshop was a source of intense pride to him, and he always enjoyed showing visitors round and explaining his plans and ambitions. Charles Darwin came to Cambridge in November 1877 to receive the degree of LL.D. and after the ceremony he and other distinguished strangers were conducted round various university buildings by members of the Museums and Lecture Rooms Syndicate. Stuart's letter to his mother tells us of his visit to the lecture room in the Jacksonian building. 'Amongst other places they came to see my museum and workshop, and I shewed them what my pupils were about, and the system I adopt for teaching them, and one thing and another of that kind, whereat they seemed greatly pleased, and more especially Darwin.' A year later Gladstone visited the new workshop. Stuart found him 'very friendly and kind' but formed the impression that he was more interested in the social aspect of the scheme of training than in the 'mechanical contrivances' which were produced for his inspection.

During the year 1878–9 further progress was made. The space originally allocated to the smithy was converted into a carpenter's shop and a new smithy built in a lean-to adjoining the main hut. Coutts Trotter, a fellow of Trinity who took a very active interest in the fortunes of the new department, gave £100 towards the cost of a machine to be chosen by Stuart, and Horace Darwin, also of Trinity, gave a lathe valued at £15. The assistant staff now consisted of an instrument maker, a fitter and a cabinet maker. Eighteen students (including six graduates) were at work during the Easter Term, and of these six intended to

[1] Preserved in the Library of the Engineering Department.

become engineers and five others to follow allied professions in which workshop training would be of direct value.

In his report to the Museums Syndicate in May 1879 Stuart warned the university that he would not be able to run his department single-handed for very much longer, and on 4 June he wrote to his mother,

I shall be very glad when I can get a holiday though I do not see my way exactly to any rest just now for my workshop greatly needs my presence always. I mean to carry it right through for a complete year and then to tell the University that there it is successfully in working order and that they must 'take it or want it or walk to the door'. It is going on and increasing and doing most successfully. I have fairly succeeded in providing a Mechanical Education in the University and a workshop where experimental apparatus of all kinds may be made for those who are to use it. But the establishment of all this is a very considerable tax on my energy.

A year later he repeated his request to the Syndicate in more urgent terms. He expressed his willingness to continue for the time being to conduct the workshop at his own risk, but asked for a competent demonstrator to assist with its management and the conduct of the classes, pointing out that his was the only department in the university involving experimental teaching in which the professor did not receive some assistance of this kind. The Syndicate supported his request, and on 15 December 1880 J. A. Fleming, B.A., of St John's College, was appointed Demonstrator with a stipend of £150 a year and the stipulation that he must not augment this by taking private pupils in any branch of mechanism or applied mechanics.

When Fleming took up his duties the Statutory Commission on university and college finance referred to in chapter 1[1] had been sitting for three years and it was becoming clear to everyone that the colleges would soon be compelled to make a contribution to the Chest from their revenues and that more money would then be available for the university to spend on its teaching staff, buildings and equipment. The need for all these was pressing, and in December 1880 the Museums and Lecture Room Syndicate obtained a grant of £1,000 from the Chest to meet some of the most urgent requirements of the Jacksonian Professor and the Professor of Mechanism. In the same month the Syndicate published in the *Reporter* a statement of needs by Stuart which

[1] See p. 24.

was in marked contrast to the courteous and somewhat diffident requests which he had made in former years.

> The increase in the number of my pupils and their growing requirements have made the premises which I now have inadequate; and it is no longer possible for me to carry on my work without some addition. I require:
>
> 1. An addition to my fitting shop. Where a number of my pupils have now advanced to making small engines and require more space for erecting them.
> 2. A room in which my pupils can be taught mechanical drawing. Hitherto I have managed this in the fitting shop, but can do so no longer owing to its being so full, and a number of my pupils requiring more elaborate drawing instruction.
> 3. An enlarged store, my present being inadequate.
> 4. Some better accommodation for a lavatory.

To meet these needs the Syndicate recommended that a new wing should be built at right angles to the existing hut and another room made by roofing over part of the space between the workshop and the porter's lodge. This was approved on 14 December and the buildings were completed before the end of March 1881. A year later a two-storeyed extension was built on to the main hut to house a gas engine to drive the larger machine tools, which now included a fourteen-foot lathe and a slotting machine.

With Fleming's help the scope of the teaching was greatly extended, and in 1881 Stuart lectured on the differential calculus and its application to mechanics and Fleming gave demonstrations and lectures on elementary applied mechanics. Fleming also took the first class in surveying and levelling, a subject which has formed part of the Cambridge engineering course ever since. In the workshops pupils assisted in the making of steam engines and scientific instruments, and in the new drawing office they were taught mechanical drawing and machine designing, graphical statics and the theory of structures.

In the Michaelmas Term 1882 Fleming was succeeded as Demonstrator by James Samuel Lyon. Lyon had an unusual background for an undergraduate of those days, having matriculated at the age of twenty-six after twelve or more years practical experience in the engineering industry. He was the son of a minister in Sheffield, but nothing more is known of his early years. He was admitted to St John's College as a pensioner in 1878, transferred to Clare in the following year and graduated as Bachelor of Arts in 1882. On taking up his duties as demonstrator in the Michaelmas Term he broke new ground by giving a course of lectures on the practice of iron and brass founding. He soon made himself indispensable to Stuart, and gained his complete confidence. With his help the old smithy was converted into a foundry, and although its smoke caused considerable annoyance to the neighbours it was considered by Stuart to be one of his most successful ventures from the educational point of view. Lyon also took charge of the workshop and supervised both the teaching and the work done to order there.

By this time the Statutory Commission had completed its work and the revised statutes came into force in 1882. A General Board of Studies was set up and Special Boards created for the separate subjects on the lines of the Board of Mathematical Studies which had been in existence since 1848.[1] The Professors of Mechanism, Chemistry and Mineralogy with the Jacksonian and Cavendish Professors were members of the Special Board of Physics and Chemistry. Stuart, who was already a member of the Council of the Senate, was also appointed to the General Board, but in comparison with the Council he thought very poorly of it.

Of all the governing bodies I have been on I never knew any which conducted its business in a more admirable manner than the Council of the University. On the other hand the General Board of Studies was about the worst, and, curiously enough, it consisted, to a very considerable extent, of the same persons. Each member of the Board of Studies was elected to champion a particular interest, the Board of Mathematical Studies, for instance, sending a representative to watch over the interests of mathematics, but the members of the Council were elected simply to govern the University properly.[2]

In the Michaelmas Term 1882 the General Board sent a questionnaire to all the Special Boards and syndicates, inviting them to state their needs for additional teaching staff, buildings, or equipment. It was clear from the replies that the increased income of the Chest due to the college contributions would be quite inadequate to meet all the demands of the departments, and in their report to the University published on

[1] See p. 25. [2] *Reminiscences* (London, 1911), p. 193.

11 June 1883 the General Board made drastic cuts in their proposals. The report recommended the establishment of a university teaching staff of readers, lecturers and demonstrators, made suggestions for new buildings, and assessed the financial implications of its proposals. An appendix contained the replies to the questionnaire sent in by the Special Boards.

Stuart's report on behalf of the Department of 'Mechanism and Applied Science' again asked the university to take over the workshop at an agreed valuation, which he thought might amount to about £2,500. He also asked for a proper teaching staff, to consist of a Superintendent of Workshops with the standing of a University Reader, three full-time demonstrators and a part-time demonstrator in surveying. He wanted the superintendent to be responsible not to the professor but to the Board of Studies for the whole conduct of the workshops, including the drawing office, for their successful management and for the teaching of the pupils in them. For the superintendent he suggested a salary of at least £400 a year together with some sum dependent on the munber of his pupils and on the success of the place, but he thought that the three demonstrators should receive no more than £50, £75 and £100 respectively, as their employment would constitute the best possible training for teachers of engineering in local schools and colleges. He also asked for a larger foundry, and suggested that the university should buy two small houses in Free School Lane adjacent to the workshop with a view to replacing them later by a permanent building to house the museum and lecture-rooms, as well as one or two small rooms for private research. In forwarding this reply to the General Board the Special Board agreed to support the appointment of a superintendent, but insisted that he should continue to be responsible to the professor. His duty would be 'to manage and superintend the workshops and the work done therein and to provide for the efficient practical teaching of the pupils, and to give assistance to the Professor in giving class instruction'. This was accepted by the Senate, the post was established on 7 February 1884, and Lyon was appointed four days later at a salary of £300 a year. Shortly afterwards Stuart appointed three of his old pupils as assistant demonstrators in the drawing office, fitting shop and foundry respectively. These were unofficial posts, without university status, paid out of the departmental income from students' fees and the profits of the workshop.[1]

Stuart also got his new buildings. An extension to the two-storey block put up in 1882 gave him two good rooms and he fitted one up as a lecture-room and transferred the bulk of his museum to the other. The university bought the Free School Lane cottages and allotted them to his department in exchange for two rooms in the main Museum block which were transferred to the Professor of Botany, and a new foundry, thirty-five feet long by twenty-five wide, was built in the garden of one of them.

During the Long Vacation of 1884 the Mechanical Workshops carried out what was probably the most remarkable operation in their history. Approval had been given to add a third storey to the building which housed the herbarium and the Department of Mineralogy, and Stuart and Lyon together evolved a scheme for jacking up the existing roof and inserting the new walls underneath it. The work was completed successfully and Stuart described the method which had been adopted in a report to the Museums Syndicate.

On each side of the principal timbers of the roof, holes were made in the walls under the wall-plates, and new timbers were slid in and placed with their ends on the two walls. It was then ascertained how much weight these new timbers would have to carry in supporting the ceiling, and what their deflection would be when their load came upon them. After being screwed down with a 'jack' to give them the required deflection, every rafter which carried the ceiling of the Mineralogical Museum was attached to them by means of hoop-iron-slings; and when the operation was complete, the nails fastening these rafters to the principal timbers

[1] Stuart's report in answer to the questionnaire contained an interesting summary of the financial position of his department, at this time.

'[Teaching] is at present done by five workmen, two lads and two boys who all belong to the category of "Assistants" and by two demonstrators, one of whom receives £150 from the University and an additional sum from me and the other of whom is entirely paid by me.

The fees of the pupils are not nearly adequate to keep up this establishment, the expenses of which are as follows:

Wages	£750
Stores and power	75
Upkeep of machinery	150
	£975

The fees of the pupils last year (exclusive of lecture fees which amounted to £40) amounted to £415, so that I had to become responsible for the deficit of £560 besides the interest of £125 on the capital involved, making in all £685. This deficit I have recovered by employing the workmen, when not employed in teaching, in working for profit, and, by this means, I have now brought the affair so that it pays the whole of the expenses except the sum which I pay to the demonstrators which is not included in the above statement.'

of the roof were very carefully cut so as to leave the ceiling hanging entirely by the new timbers. So nicely was the deflection calculated, that no timber moved upwards or downwards more than the eighth of an inch when the load came upon it. Next, an iron beam on either side of the principal was slid under the wall-plate, and these also, six and twenty in number, were made to rest on the two walls, which have a clear span of twenty-five feet. The lifting of the roof then began. It was accomplished by means of twelve screw-jacks, one of which was placed under the end of each of the principal timbers on the same wall. Any straining of the roof was prevented by working all the jacks simultaneously, the men giving each jack half a turn at the word of command of the Superintendent. All the principal ends being thus lifted one foot, packing was put across the two iron beams on either side of each principal, and the principal was lowered onto it; the jacks were then all taken to the other side of the building and a similar process was carried out there. Brick pillars were built temporarily on the iron beams to carry the principals while the roof was being lifted the necessary twelve and a half feet so as to leave the walls of the new storey perfectly free from strain while they were being built. Finally, when the walls were complete, the roof, which had been lifted a few inches too high, was lowered on to them and made secure. Not a single slate or nail in the roof was broken or strained, and the whole work of lifting the roof and building the exterior walls was completed in seventeen working days, the weight lifted being fifty tons, and the total length of roof one hundred and ten feet.

...The whole affair was planned and executed by Mr Lyon, to whose untiring superintendence, and practical and theoretical skill, the success of the undertaking is entirely due...The work has been very useful for such of my pupils as were present during the Long Vacation, and I consider that this has been one of the chief advantages to the University of the undertaking. It afforded the opportunity for the execution of detailed drawings of a valuable educational kind, and for my pupils not only becoming acquainted with unusual operations, but also—what I consider equally valuable—seeing how ordinary building and other work is carried out.

The thanks of the university were conveyed to the Department of Mechanism for the care, skill and economy with which the building operations were conducted, but in the discussion of the Museums Syndicate's report some doubts were expressed about the wisdom of employing a university department on work of this kind. However, the general feeling of the Senate was that the risk had been justified by the success and economy of the operation.

Stuart was now at the peak of his career in the university. He was a member of the Council of the Senate and the General Board of Studies, of the Special Boards of Mathematics and of Physics and Chemistry, of the Syndicates for the Museums and Lecture Rooms, Local Examinations and Lectures and the Training of Teachers and an Elector for the Professorships of Physics and Political Economy. During the last two years the number of his pupils had doubled, and nearly seventy were now attending the course in the workshops. These were now housed in modern, though temporary, buildings, and in the same period over a thousand pounds worth of new equipment had been purchased for them, almost entirely with Stuart's money. In addition to the regular courses of instruction given by Stuart and his staff, practising engineers were brought to in give lectures on such subjects as bridge designing, the foundations of viaducts and piers, tunnelling and sewage engineering. Examination results were on the whole satisfactory, particularly in those subjects which had a practical application, and the reputation of the department stood high, both inside the university and in the industrial areas where Stuart had most of his contacts with the engineering profession. All this was the creation of one man, and nothing like it had ever been seen in Cambridge before, but the department was not yet integrated into the rest of the university and its weakness was revealed when Stuart's interest began to flag and his control relaxed. The turning point came in November 1884 when he was elected member of Parliament for the Borough of Hackney.

MOUNTING TENSION, 1884–1887

The election campaign at Hackney was not Stuart's first attempt to enter Parliament. He had stood as a Liberal candidate when one of the university seats fell vacant in November 1882, but the radical views which he had inherited from his father and the active support which he gave to Mrs Josephine Butler in her campaign for the rights of women were not popular in the university and he was defeated by 3,491 votes against 1,301.

His conduct of the campaign at Hackney aroused great resentment among some of his university colleagues, and it may be that the death of his mother in 1882 had removed a restraining influence on his impetuosity. Professor G. F. Browne, who as Secretary of Local Examinations had worked closely with him in establishing the extra-mural lectures, expressed his feelings in a vitriolic letter to the *Cambridge Chronicle*.

Professor Stuart's address and speeches at Hackney are a terrible blow to the reputation of those who pressed him upon the University constituency two years ago...Their academic and cultured candidate...this chosen friend of Bishops and Deans and Masters of Colleges, goes to Hackney and pours forth a flood of ultra-Radicalism, of crazy fads, of coarse misrepresentation and vulgar abuse, which must make his Cambridge friends shudder.[1]

Stuart won the seat at Hackney by a comfortable majority, but he did not hold it for long. There was a general election in the following year and another in 1886, and in both he successfully contested the Hoxton division of Shoreditch, a constituency which he represented for sixteen years.

As his preoccupation with politics increased he became more anxious than ever to free himself from responsibility for the workshops, and in his annual report for 1883–4 he repeated his request that the university should buy his plant and equipment and transfer the control of the workshops to the superintendent. In the autumn of 1885 the Council of the Senate appointed a committee to look into the problem, and their report was published on 25 January 1886. It reviewed the history of the workshops from their inception and described exactly how they had been equipped and financed. It expressed the view that while the venture, which was quite without precedent, was in the experimental stage it could only have been carried out at the risk and on the responsibility of the Professor, but that stage was now over; the workshops had justified themselves and the present anomalous situation ought to be rectified as soon as possible. The take-over should not present much difficulty, as careful accounts of receipts and expenditure had been kept from the beginning and the machinery had been maintained in admirable order. The committee recommended that all Stuart's property should be purchased at an agreed valuation and that the workshops should then be run as a university department on the same lines as at present.

The report was discussed on 4 February, and its reception was generally favourable, though James Mayo of Trinity, who was just

[1] *Cambridge Chronicle*, 15 November 1884.

entering on a long career as self-appointed watch-dog of the university finances, thought it wrong that the Old University should be called on to pay for the new one which was rising up. Support for the proposals came from Coutts Trotter, John Willis Clark[1] and several others. Dr Phear declared that the department had enhanced the reputation of the university and increased the number of its students by attracting men who would not have been interested in the old curriculum. The report was approved without opposition on 11 March.

On 10 May the Council published a further report, the first part of which informed the university that Sir Frederick Bramwell, President of the Institution of Civil Engineers, had at the request of the Vice-Chancellor nominated a London firm of chartered accountants to act as valuers. They had assessed the workshop machinery, tools, stores, book debts and work in progress 'as a going concern between parties' at £4,019. 12*s.* 0*d.* and the Council recommended that the university should pay this sum to Stuart and take over the workshops as from 16 April 1886. The second part of the report contained draft regulations under which the workshops were to be managed by the superintendent under the direction of the professor, while the Museums and Lecture Rooms Syndicate continued to be responsible for the maintenance of the buildings. Payments to be made out of Departmental receipts were listed under four heads:

(*a*) Wages, tradesmens bills, general maintenance expenses including repairs to machinery, and payments to Assistant Demonstrators.

(*b*) 10% of the value of the machinery and tools to go to a depreciation fund.

(*c*) 3½% on the capital invested to be paid into the Chest.

(*d*) The balance to be carried to a reserve fund under University control.

The Professor was to report annually to the Senate on the state of the workshops, and the Financial Board was to arrange for an annual audit of the departmental accounts and for periodic valuations of the machinery and plant. The capital invested was to be raised to £4,400 by the addition of £130. 8*s.* 0*d.* to the price fixed by the valuers.

[1] John Willis Clark was the nephew of Robert Willis, the Jacksonian Professor. He was secretary of the Museums and Lecture Rooms Syndicate from 1880 to 1891 and Registrary from 1891 until a month before his death in 1910. He inherited Scroope House from his father, Professor William Clark, and lived there from 1885. He was joint author with Willis of *The Architecture of the University and Colleges of Cambridge* (1886).

This report aroused serious misgivings, particularly among those members of the Senate who disapproved of engineering as a university subject and those who thought it wrong for a department to indulge in private trading. Others had doubts about the financial arrangements suggested by the Council and wanted further information. George Darwin, the Plumian Professor, presented a memorial protesting against the way the valuation had been conducted and suggesting that the inventory included gifts to the department as well as purchases by Stuart.

The discussion on 5 May opened amicably with some criticisms which were answered on behalf of the Council by Coutts Trotter. The real attack was then launched by T. F. C. Huddleston of Trinity, who was known to be a strong opponent of the teaching of engineering in the university. By quoting detailed figures he tried to prove that the proposed arrangements gave no proper safeguard for the capital which would be invested in the workshops, and he asked for a great deal more information. Stuart had evidently been forewarned of what was coming and he replied by reading a long and detailed apologia. His whole attitude was one of injured innocence, and it was evident that he had been deeply hurt by aspersions against his integrity. He had received a letter from 'an intimate and honoured friend' who told him it was being said that he had 'made a good thing out of the workshop' by putting into it the fees of the students received on a tacit understanding that they would go towards the maintenance and not the expansion of the department, thus increasing out of the fees the value of what the university would take over. He strongly contested this accusation, and gave figures to refute it, but complained that after serving the university for twenty years it was nothing short of a shame that he should be called upon to make this statement. He ended by saying that he did not wish to remain responsible for what was not his statutable duty, and although he would always give his deepest care to the interests of the workshops he wished the responsibility for them to be transferred entirely to the superintendent. This statement clearly caused some consternation in the Senate House and Browne at once rose to propose the adjournment of the discussion to allow the Senate time to consider the 'clear, interesting and in some ways remarkable speech of Professor

Stuart', which was agreed to after the motion had been seconded by one of the signatories of Darwin's memorial.

The discussion was resumed on 12 June and was opened by Darwin, who explained the misgivings which had inspired the memorialists and recounted the steps which had been taken to rectify the mistakes in the inventory about which they had complained. He blamed the Council for having failed to ensure that the valuation was accurate, but could not see why Stuart should feel shame because, as the vendor of valuable property, he was asked questions about it by the prospective purchaser. He was followed by other speakers, some favourable and some hostile to the report, and Stuart rose seven times to answer questions or reply to criticisms. His attitude was defensive and irritable, and it was clear that his nerves were on edge. Lyon was angry and even more aggressive. He reacted strongly to a suggestion by J. W. L. Glaisher, Tutor of Trinity College, that there should be a special Syndicate of Management for the workshops similar to those for the Library and the Press, saying that there were very few gentlemen in the university competent to sit on such a body, except as a mere formality, and he thought it not well that members of a syndicate should be there only to vote on things they did not understand. Coutts Trotter as usual tried to restore calm and common sense to the discussion. No one was satisfied with the state of technical education in England, but students in the great engineering works got little theoretical training, and many practical engineers supported Stuart in the belief that much could be taught in Cambridge that could not be taught elsewhere. Glaisher, who was one of the severest critics of the workshop courses, spoke highly of the lectures, and on this congratulatory note the discussion ended.

No further action was taken during the Long Vacation, but an amended report was issued on 13 December. No major changes were made in it as a result of the discussion, but some items which had not been bought by Stuart were removed from the inventory, and the capital sum on which interest was to be charged was reduced to £4,100. Stuart was to be paid £3,975. 8*s*. 0*d*. for the equipment which was his property, and in future the superintendent would receive one-quarter of the fees paid by students in the workshop in addition to his stipend. A syndicate composed of the Vice-Chancellor or his deputy, the Professor of Mechanism and three other members of the Senate would report annually on the condition and progress of the workshops and publish an abstract of the departmental accounts after they had been approved by the Financial Board. Thus Stuart's request to be relieved of all responsibility and Lyon's objection to a syndicate of management were both overruled. The report was approved without opposition on 21 January 1887.

While the purchase of the workshops was being negotiated Stuart worked for the institution of an honours examination for his abler students. The 'Special' in Mechanism and Applied Science was much more difficult than the other examinations for the ordinary degree, but it lacked the prestige of a tripos and was not up to the accepted intellectual standard of an honours examination at Cambridge. The best students were therefore compelled, if they wanted an honours degree, to sit for the Mathematical or Natural Sciences Tripos, and if they intended to become engineers this meant that they had to devote part of their time to subjects outside the range of their main professional interests. It also meant that, as undergraduate teaching is inevitably directed towards the degree examination, no comprehensive honours course could be organized for intending engineers. Somewhat similar difficulties were experienced by the best students of physics and chemistry. They could obtain honours in the Natural Sciences Tripos, but this was non-mathematical and not very highly regarded, while the Mathematical Tripos, which continued to attract the élite of the undergraduates, was too specialized to meet the requirements of scientists. Stuart therefore had little difficulty in persuading the Special Board of Physics and Chemistry that a new type of honours examination was needed, and its proposals were put before the Senate in a report dated 14 December 1885.

The suggested examination was intended primarily to test the proficiency of students of engineering, but the interests of men reading physics and chemistry were also to be catered for. No changes were proposed in the existing Natural Sciences Tripos, but as an alternative to Part I eight papers were to be set on the mathematics and mechanics required by engineers and on the elements of physics and chemistry. A practical examination in engineering would be added to those already set in physics and chemistry and candidates would be required to pass in at least one of the three subjects. A man who, in his third year, passed both the written and practical examinations with credit would be entitled to a degree with honours. A new examination was also proposed as an alternative to Part II of the Natural Sciences Tripos. Six papers would be set in (*a*) Engineering, (*b*) Physics, and (*c*) Chemistry, and there would be three practical examinations. Candidates would be allowed to take one or more subjects and again all who passed with credit would obtain an honours degree.

The discussion took place on 4 February and was chiefly remarkable for the speeches of Coutts Trotter and Stuart. Mayo, who opened the discussion, thought that the proposed examination would only constitute a minor Mathematical Tripos, and though it might convince the outside world that Cambridge engineers knew some mathematics it would convince no one that Cambridge mathematicians knew any engineering. The mere existence of Stuart's excellent course was all that was needed. Coutts Trotter explained that a new examination with a greater mathematical content was required by both engineers and physicists, as the puzzles and mental gymnastics of the Mathematical Tripos did not meet their requirements. Most engineers needed much more mathematical training, but they could not afford to spend three, or even two, years working for a tripos which involved a great deal of mathematics of no practical use to them. Conversely a course of mathematics suitable for an engineer or physicist would not help a man to a high place in the tripos, and an entirely new examination was therefore essential. As things were, the best men reading engineering and physics would not be content with third class honours, which was all they could expect if they devoted the time and attention needed for a proper grasp of their own subject.

To a speaker who asked how the new Second Examination in Physics and Chemistry would differ from the existing Part II of the Natural Sciences Tripos Stuart replied that the Natural Sciences Tripos excluded mathematical treatment and was not good enough for men who wanted to study physics mathematically. He went on to explain in detail how the mathematics required for the various papers in the new examination differed from that required for the tripos. He always advised his best students to take the tripos, but their natural desire for a good place inevitably led them to work hardest at the trick questions, not at the mathematics they would need as engineers. The proposed examinations would enable him to adapt his teaching to the real requirements of engineers and thus make the Cambridge course the best in the world. The practical examination, too, would have a most important effect, as it would lead the students to embark on practical work early and combine it with their theoretical studies instead of leaving it until after they had taken the tripos. A new examination was also needed for men who wished to read mathematics not as an ultimate study but as a help and means to other studies. Those who wished, as he himself wished most ardently, to retain the study of mathematics at Cambridge as an education of the mind ought to be prepared to see a selection of the earlier branches of it adapted to the needs of other fields of human knowledge.

Lyon, as a practical engineer, agreed that it was impossible for a man who was reading seriously for the Mathematical Tripos to get a worthwhile knowledge of engineering in the time available to him. The university would never be able to send out a man stamped as an engineer in the sense in which they sent out a man stamped as a doctor, until they had established an examination designed for the purpose. Other speakers suggested that the Mathematical Tripos might be modified to meet the needs of the scientists, but Dr N. M. Ferrers, Master of Caius, concluded the discussion by begging the Board to stand their ground. If they tried to adapt the Mathematical Tripos to meet the needs of the engineers they would get one bad examination in place of two good ones. At a congregation on 11 March the Senate accepted this point of view, and the report was approved without a division.

The regulations for the new examination, which was to be known as the Mechanical Sciences Tripos, were published in a report by the Special Board of Physics and Chemistry on 1 May. The preamble stated that, as the university ought not to examine in subjects for which it did not supply adequate teaching, questions on engineering would be confined to mechanical engineering and such elementary parts of civil engineering as could be taught in Cambridge. The schedule followed the lines of the report which had already been approved. In Part I the written examination was to occupy five days and the practical three. Part II would extend over nine days, of which the first three would be occupied by papers on engineering, physics and chemistry set simultaneously and the last six by the practical examinations, two days to each subject.

The report was not discussed until 19 October, after the Long Vacation. It was introduced by the Vice-Chancellor, Dr C. A. Swainson of Christ's, who strongly supported the proposals. He told the Senate that during the summer he had received assurances from several quarters of the high character assigned to the teaching of engineering in the University. He had been informed that young men were sent to Cambridge because they could get better teaching in the Department of Mechanism than anywhere else in the country. Coutts Trotter stressed the danger of overloading the syllabus, and suggested that the physics paper should be optional for engineers, but J. J. Thomson, the Cavendish Professor, believed a knowledge of physics to be an essential part of the scientific training of engineers—even more necessary, perhaps, than workshop practice, which could be obtained after leaving Cambridge. This challenge brought Stuart to his feet. Where Engineering Schools had failed it was because they had divorced the theoretical work of mathematics from the practical work of good engineering business. In his opinion the essentials of an engineering education were mathematics, workshop practice, drawing and designing. If electricity, magnetism and chemistry were taught they should be included in the latter part of the course and examined in Part II, but, whatever the content, an engineering tripos was desirable both from the student's and from the university's point of view. The proposed schedule was intended to attract a new type of student, but these men would not come to Cambridge if their workshop apprenticeship had to be postponed until after graduation, as it would then be too late for them to enter the profession. W. H. Macaulay of King's wanted a more scientific type of training in what might be called an 'engineering laboratory', but Lyon thought that whereas engineers had formerly had too little theory in their training the pendulum was now tending to swing too far in the opposite direction. He thought the scientific value of workshop training was underrated, but it must be given while the student was still young. Another speaker stressed the importance of practical work when studying a practical subject, and believed that in this case the theory should cover the field common to engineers, physicists and chemists while each did the practical work in his own subject. R. T. Glazebrook of Trinity[1] agreed with the importance of practical work but thought that the measurement of physical properties could be carried out better in the Cavendish Laboratory than in the Mechanical Workshops.

As a result of this discussion an amended report was published on 5 February 1887, but the only change of importance was in the regulations for the practical examinations in Part I, in which candidates were now only to be required to take one of the three subjects—engineering, physics or chemistry—instead of two. This alteration was not considered of sufficient importance to justify another discussion, and the Grace for approving the report was submitted on 10 March. No previous notice of non-placet is recorded and there is no evidence of the publication of fly-sheets, but the report was rejected by the Senate by 73 votes to 59.

On the surviving evidence it is difficult to account for the rejection of the tripos when so much had been said in its favour and so little opposition had been expressed to the idea of making engineering the subject of an honours degree. It seems to have been more in the nature of a vote of no confidence in Stuart himself than an indication of hostility towards the subject which he taught. His political activities had alienated many of his former well-wishers, and he had been accused of 'making a good thing out of the workshops'. It was unfortunate that the proposals for the purchase of the workshop plant and the establishment of the tripos came up together at a time when he had offended so many members of the Senate. The defeat must have been a great disappointment to him, but in his *Reminiscences* he speaks of the affair without any trace of bitterness.

There was a great deal of opposition to that Tripos when I first proposed it towards the latter part of my tenure of office. The Tripos consisted of a union of practical engineering with the application of mathematics in so far as it was needful for such purpose, and I always argued that the selection of mathematical

[1] Later the first Director of the National Physical Laboratory.

studies necessary for that purpose offered a most admirable mathematical education to those who might take it up. The opposition was principally based on a recrudescence of the original opposition which I had to overcome in connection with my Workshop, namely that anything that was not purely abstract in its character was unsuitable for a University.[1]

This latter statement was undoubtedly true as far as some people were concerned, but there were others, including Macaulay, Glazebrook and Darwin, who wanted engineering to be taught in the university, but believed that Stuart was teaching it in the wrong way.

THE BREAK, 1887–1889

The transfer of the workshops to university ownership and the rejection of the tripos opened a new phase in the history of the department, but it was not a happy one for Stuart. Work went on much as before, but there were no new developments and it is clear that Stuart's delegation of authority was beginning to result in a general slackening of effort. The examiners' reports on the 'Specials' for 1887 and 1888 make depressing reading.

The style of the answers was often bad. The candidates showed very little knowledge of what to aim at in answering questions; and consequently the answers were incomplete and badly put...The practical work on the whole was satisfactory, but the candidates did not handle the tools in a workmanlike manner, nor did they shew skill in putting them into working order...

One of the candidates distinguished himself in the examination in workshop practice, but with that exception the marks obtained throughout the examination barely exceeded the minimum required for passing, and the standard of proficiency was far short of what might fairly be hoped for from engineering students before entering upon professional work.

About this time an incident occurred which, though comparatively trivial in itself, had serious repercussions. In October 1886 the Professor of Chemistry, G. D. Liveing, obtained approval for the fittings for his new laboratory to be made in the Mechanical Workshops. Lyon undertook to do the work for about £1,800 and the Financial Board allocated £2,000 for the purpose. In May 1888 the Board published a report stating that although all the money had been spent much work still remained to be done. Lyon now estimated the total cost, including

[1] *Reminiscences*, p. 182.

15 per cent commission for the department, at not less than £4,350. Stuart had been asked for an explanation, but he disclaimed all responsibility and referred them to the Superintendent of the Workshops. Lyon, in a letter appended to the report, said that his original estimate was no more than a rough guess based on inadequate information.

Opening the discussion on the report on 31 May on behalf of the Financial Board, Dr Porter, Master of Peterhouse, regretted the mistake in the estimates, but said that he had himself inspected the completed fittings and considered the workmanship to be of a very high order, reflecting great credit on the department. W. Cunningham of Trinity thought it regrettable that personal considerations were allowed to intrude whenever matters affecting the Department of Mechanism were discussed and wished to see the whole question of the department raised before the Senate was called upon to vote on the report. W. E. Heitland of St John's agreed, and went so far as to say that if the continued existence of the department were put before the university now, and not allowed to stand over till the winter when the Senate would have forgotten what had taken place, there would be a large majority in favour of its total suppression. Another speaker said he had heard Lyon's argument that it was impossible to make estimates used before, and it had been in the bankruptcy court. Huddleston asked whether the work had been of any educational value to the students, and Lyon replied that it had not been of the slightest value to any of them. Tempers on all sides were evidently near breaking point, but Porter and others did what they could to ease the tension and the discussion ended inconclusively.

The report was approved without a division on 7 June, and the Financial Board was authorized to have the work completed during the Long Vacation at a further cost of £1,100. The final cost in fact amounted to £5,107. 16*s*. 2*d*. and in February 1889 the Financial Board had to come back to the Senate a third time to obtain formal approval for the additional expenditure.

The Mechanical Workshops Syndicate, which had been appointed in January 1887 when the university took possession of Stuart's plant and equipment, issued its first report on 12 March 1888. Lyon's accounts showed a great falling off in the work done for other departments, and the Syndicate pointed out that if the trend continued it would be impossible to pay the interest on the invested capital as required by the regulations. There was also a drop of £80 in students' fees for the Michaelmas Term 1887 compared with the previous year due to the fact that when the Engineering Tripos was rejected some of the best students immediately withdrew from the workshop course. The only satisfactory piece of information which the Syndicate could offer was the award of a silver medal to Lyon for an improved screw-cutting lathe designed and made in the workshops.

More than a year passed, and the fortunes of the workshops declined still further. The Syndicate did not issue its second report until 15 June 1889 after many weeks had been consumed in wrangling and recrimination. When it eventually appeared it was a colourless document, giving little indication of the true state of affairs in the department. It was in two parts, the first containing a survey of the work carried out during the year and the second giving a very guarded version of the financial difficulties which had been encountered.

The accounts for 1887 had been passed to the Financial Board early in the new year, but the form in which they were presented was considered so unsatisfactory that a firm of chartered accountants had been asked to frame a system for use in the future. This system, however, had been found 'impracticable', and the accountants were then asked to prepare a balance sheet and profit and loss account themselves. They did so, but the Syndicate objected to certain entries and the Financial Board ultimately adopted an amended version which was appended to the report. It showed a loss of just over £94 in 1887 and a profit of £178. 10*s*. 6*d*. for 1888, but the latter included a commission of £420. 12*s*. 9*d*. from the Chemical Laboratory, which would certainly not be repeated. The Financial Board were strongly of the opinion that the accounts ought to be kept in accordance with the recommendations of the accountants, and that for this purpose adequate assistance should be given to the superintendent, but the syndicate, of which Stuart was a member and Lyon secretary, refused to agree. They did not see how the additional expense of providing clerical assistance could be met, nor did they think that the expense would be accompanied by adequate advantage. The report concluded with a recommendation, not supported by the Vice-Chancellor (who was also, of course, Chairman of the Financial Board) that if new financial regulations were drawn up as the Board had requested interest on the capital for the years 1889, 1890 and 1891 should not be paid into the Chest but carried to a suspense account.

The report was signed by C. E. Searle, the Vice-Chancellor, and all the members of the Syndicate—A. Cayley, the Sadleirian Professor, Stuart and W. N. Shaw, who had taken the place of Dr Phear. It made no mention of the report on the workshop accounts submitted by the auditors, but this had been seen by the Financial Board and when the Syndicates' report came up for discussion its contents were known or suspected by many other members of the Senate. A copy of it has been preserved in the University Archives. It reads as follows:

> We regret to say that the figures in the accounts of the Department at 31st December, 1888, submitted to us were in so many respects incorrect that we have found it necessary to prepare a fresh statement of accounts... We found that the figures of the accounts at 31st December 1887, in many respects required adjustment... As regards the balance sheet at 31st December, 1888, submitted to us the figures were wrong in many respects...
>
> (*a*) The Cash balance was understated.
>
> (*b*) The Plant account was overstated.
>
> (*c*) The Debtor's account was overstated.
>
> (*d*) The Balance of the Surplus and Reserve account was not properly stated.
>
> (*e*) The total of the Creditors was not properly stated.
>
> (*f*) The figures of the accounts, submitted to us were in some cases not recorded in the ledger kept at the Workshop and many items had not been posted.
>
> (*g*) There was no check upon the correctness of the book-keeping in consequence of no Profit and Loss account having been prepared...

It is difficult to resist the conclusion that the auditors believed Lyon to have 'cooked' the accounts in order to disguise the true state of the departmental finances, and some members of the Financial Board undoubtedly held this view. Lyon contested both the accuracy and the implications of the auditors' statement, and the surprising thing is that Stuart, who had himself kept the early workshop accounts with such meticulous care, supported him. After weeks of negotiation between the Syndicate and the Board the compromise report of 15 June was drawn up by Professor Cayley and accepted by both parties.

The discussion on 24 October was opened by Huddleston, who complained that no report or minute of the Financial Board had been published, and asked that the auditors' statement should also be placed before the Senate. Until this was done he could not even discuss the recommendation to waive the payment of interest, but he suspected the department to be bankrupt. He asked how many pupils were now in attendance, and how many of them were not members of the university. The General Board and the Council had recently pointed out that the young men in the town ought not to be allowed to attend university lectures without first becoming undergraduates. He expressed general dissatisfaction with the state of the department, and thought its whole position ought to be reviewed by the Senate.

Dr Porter agreed that the auditors' report should have been published. He had been a member of the committee of the Financial Board which had dealt with the matter, and the accounts were the worst he had ever seen. They were in such a state of confusion that the accountants had charged over £100 for going into them and drawing up a balance sheet. In spite of this the Syndicate had refused to agree to the appointment of a clerk to help the superintendent with the accounts, and he was convinced that nothing but a very strong expression of opinion on the part of the Senate would ever get them kept in a proper manner. He asked that a new Syndicate should be appointed to inquire whether the technical instruction given in the workshops was worthy of being continued in the university. This had already been queried in a number of fly-sheets, and it might be difficult to obtain an impartial body, but he believed that some of the eminent civil and mechanical engineers who were members of the university would be willing to give advice on the matter. If it were decided to continue the workshop instruction the next question to be answered was what was its value to students intended for the engineering profession. On this point there were decided differences of opinion, but if the answer to both these questions was favourable to the continuance of the workshops he would cheerfully agree to the surrender of income, for that was a small point compared with the others he had suggested.

Comparatively minor matters were raised by other speakers, but the main contribution to the debate came from Stuart himself. The tone of his speech was dignified, reasonable and conciliatory, and he showed no signs of the irritability which had been evident on some previous occasions. He began by answering some of the questions which had been asked and then proceeded to a general defence of his conduct of the professorship. In reply to Huddleston he said there were at present about thirty-five pupils, of whom nine were not members of the university. As regards the payment of interest he quoted Coutts Trotter as saying in 1886, 'If the Department was to be as prosperous as it was at present, it was desirable to do rather more work for other departments than latterly.' And again, 'A little more work given to the department would make it hold its own in the future as heretofore, and would considerably increase the usefulness of the department. It would be an act of sheer folly to stop the system.' He admitted that the decrease in work for other departments was due to doubts as to the reliability of the estimates, but those given to Chemistry had not claimed to be accurate estimates framed on proper information, and this was the only case where they had not been well within the mark. The workshop accounts appeared to him, as a member of the Syndicate, to be quite adequate, but if the Financial Board wanted a different system their wishes must be complied with. It would be an expense to engage a book-keeper, who would certainly not have enough work to do, but perhaps some local accountant could be employed part-time.

The question had been asked, could the lectures go on as well without the workshops? This was precisely the question he had to answer when he was elected to the professorship. Many had hoped that his appointment might be the means of bringing the university into practical contact with the engineering profession, and he had come to the conclusion that that could only be done by providing the handicraft training which the large majority of engineers were obliged to get at an age earlier than that at which they would be leaving the university. He also felt that lectures which were not purely abstract should be given to men who were not wholly ignorant of the mechanism and use of things treated of. He would have acted differently in places where there was an engineering industry, but in Cambridge the workshops were essential if the course was to touch more than the mere fringe of the profession. The university could re-examine the problem and was entitled to change its mind, but if it decided that the department was not worth going on with a great gift to the university would be rejected. As for the criticism that the work done for the Chemistry Laboratory had had no educational value, he reminded the Senate that the workshops had a dual origin. It was the Board of Mathematics which had expressed the desire for a department capable of providing special equipment made under supervision, and he had provided the facility quite independently of the teaching given in the same building.

In conclusion, he hoped that the Senate would think very seriously before abolishing all that had been built up, and if they did so they must consider how they could keep hold of the engineering profession without it. He believed that he had extended the usefulness of the chair he held, but if they could get better advice from others, let them take it and give this up; but looking at all the facts he for one was satisfied with the results so far achieved.

None of the speakers who followed Stuart showed much sympathy for him. W. J. Lewis, the Professor of Mineralogy, said that he had stopped giving work to the department because he could get it done equally well elsewhere for a half or a third of the cost. He too asked for a strong syndicate to look into the whole question of the workshops and the teaching. He believed that the want of success of the department was due to two fundamental errors—the time spent on acquiring mechanical skill which should have been spent on learning the principles of engineering and the attempt to combine manufacturing with teaching. He also wanted to see the various documents dealing with the accounts, and thought that even if it would not be wise to publish them they should at least be on the Registrary's table for inspection by members of the Senate.

Other speakers complained about the accounts, and Lyon then rose to defend himself against the charge of contumacy. He said he had stated repeatedly that it was impossible for him to keep the accounts as they would be kept by a professed accountant—he had neither the competence, nor the inclination, nor the time to do so. He reminded his listeners that until the department was taken over by the university Stuart had kept the accounts with the help of a boy and conducted all the correspondence both on business matters and with parents and guardians. All this had devolved on him, as well as the management of the workshop, and as he had teaching duties in addition he had reduced the accounts to the simplest possible form in order to save time.

The argument went on for some time longer, and finally Dr Porter wound up by saying bluntly that he had never looked on it as Lyon's work to keep the accounts, which was tantamount to saying that the responsibility was Stuart's.

The recommendation in the report that the payment of interest should be waived for three years was never submitted to the Senate for approval. Instead, on 5 December a new syndicate was appointed

> to consider the regulations under which the Mechanical Workshops are at present managed and carried on, their financial condition and prospects, and their value as a preparation for the profession of an engineer and for students in the University other than those preparing for this profession; to confer with Professor Stuart; to consult at their discretion with other persons; and to report to the Senate on the whole question of the Workshops before the division of the Easter Term, 1890.

Six days later Stuart wrote to the Vice-Chancellor resigning his professorship. His letter and the Vice-Chancellor's reply were published in the *Reporter* on 11 February.

> Queen Anne's Mansions
> St James' Park, S.W.
> December 11th 1889
>
> Dear Mr Vice-Chancellor,
>
> I beg leave to inform you that, owing to the increasing demands of my political and public engagements, it is my intention to resign the Professorship of Mechanism and Applied Mechanics before the end of the present academic year.
>
> I had till lately hoped to be able to retain the office until I should have seen established an engineering laboratory and an Engineering Tripos, so as to complete the contribution I have endeavoured to make to Engineering Education in the University. But the terms of appointment of the Syndicate just elected seem to me to postpone that possibility, as it is not to enquire into these matters but into the value as an element of Engineering Education of the existing workshops.
>
> The University has of course an undoubted right to enquire into that as into every other part of its own machinery of education; but, though I have been for some years dissociated from the direct management of the workshops, under the existing regulations, I cannot doubt that the strong view held by me concerning

them would, if urged by me as Professor, have a preponderating weight with any responsible body of persons in the University. This would be as it ought to be, if I contemplated continuing to retain the Professorship for any length of time; but as I do not contemplate that, it seems to me desirable that the University in settling this matter, which respects the future, should be able to do so entirely on the merits of the question itself, uninfluenced by unnecessary regard for any personality and without being in any doubt as to my intentions.

I believe therefore that I best consult the interests of the University by thus at once making my intentions known, and at the same time by not formally resigning until later in the academic year, so that the University may, if it desires, have the opportunity of making up its mind before the election of my successor as to the machinery of education which it wishes to put at his disposal.

I am yours faithfully,
James Stuart

P.S. Will you kindly acknowledge receipt of this letter and also kindly bring it before the proper bodies.

Trinity Lodge,
Cambridge,
December 17th, 1889

Dear Professor Stuart,

I received your letter of the 11th with much regret.

Your services in the cause of Applied Mechanics in the University have been so signal, and the part you have played in bringing the name and power of the University before the minds and hearts of the masses of the people has been so conspicuous, that when you leave us, we must necessarily feel the void.

I have no right to ask you to re-consider your decision which has clearly been made after serious deliberation, but I confess I should be glad if I were allowed to bring your letter before the Council of the Senate before your intentions were publicly announced.

We shall meet again, if all be well, on Monday, January 13th. Please let me know if I can till then regard your letter as confidential.

Believe me to be most truly yours,
H. Montague Butler,
Vice-Chancellor

Stuart replied on the following day.

Dear Mr Vice-Chancellor,

I shall be glad to fall in with the suggestion of your letter of the 17th, and agree to regard my correspondence with you on the matter of my Professorship as in the meantime confidential.

I thank you for your kind expressions.
I am yours faithfully,
James Stuart

THE MECHANICAL WORKSHOPS ENQUIRY SYNDICATE, 1890

The Syndicate appointed on 5 December 1889 to report on the financial position of the workshops and their educational value consisted of the Vice-Chancellor and twelve other members of the Senate who were known to be interested in the problems of engineering education.[1] Among them were Osborne Reynolds of Queens', who had been Professor of Engineering at Owen's College since 1868, R. T. Glazebrook of Trinity and John Hopkinson of the same college, an eminent electrical engineer of whom more will be said later.

When the Syndicate learnt of Stuart's impending resignation they invited him to a conference and after hearing his views decided that instead of limiting themselves to the terms of their appointment it would be better to extend the scope of their inquiries to cover the whole question of the future development of the Engineering School at Cambridge. The recent disputes had shown that there was some opposition to its continuance in any form, but the Syndicate were unanimously of the opinion that it should continue and that it should be equipped as completely as experience showed to be practicable. These views were put to the Council of the Senate in a memorandum to which the Council appended a recommendation that the Syndicate should receive further powers 'to enquire whether it be desirable to develop further the Engineering School in the university on the lines suggested in the memorandum and the probable cost of any such development'. Six members of the Council, including, surprisingly, Henry Latham, who was now Master of Trinity Hall, refused to sign the report, which was brought up for discussion on 22 May 1890.

Several speakers disagreed with the views put forward in the memorandum and opposed the recommendation to extend the powers of the Syndicate. Among them was Heitland, who pointed out that if a real Engineering School were established it would have to be maintained in first rate order, and this would involve additional cost at a time when the university was unable to meet the needs of existing

[1] H. Montague Butler, Trinity, Vice-Chancellor; Dr J. Porter, Peterhouse; Professor H. Sidgwick, Trinity; Dr J. W. L. Glaisher, Trinity; Professor A. Cayley, Trinity; Professor G. F. Browne, St Catharine's; G. B. Atkinson, Trinity Hall; O. Reynolds, Queens'; R. T. Wright, Christ's; J. Hopkinson, Trinity; J. N. Langley, Trinity; R. T. Glazebrook, Trinity; A. R. Forsyth, Trinity.

departments. He thought it would be better to wait until the general situation improved, when the proposal might be reconsidered. Glazebrook, on the other hand, supported the recommendation on the ground that it was impossible to decide how the workshops would fit into a scheme in which they might be subordinate to an engineering laboratory until the nature of the Engineering School itself had been determined. Dr Porter said that since he became a member of the Syndicate he had taken every possible opportunity of visiting engineering schools in England, and had come to the conclusion that the education in them was of the highest possible value, second only to that in the Cavendish Laboratory. He referred to the Engineering Laboratory which had been established by Professor Kennedy in University College London, and hoped to see a similar one in Cambridge which would grow from small beginnings under a professor who would have his heart in his work. This virtually ended the discussion, but before the report was submitted to the Senate notice of non-placet was given by R. S. Scott of St John's, one of the Council members who had refused to sign it. This was followed by the publication of a number of fly-sheets, several of which are of considerable interest.[1]

The first, which was dated 31 May, was by Osborne Reynolds, who offered his views as one who had long devoted attention to the development of Schools of Engineering and had become convinced that an engineering school was now essential to the completeness of a university. The basis of engineering, he argued, was Applied Mechanics, and this demanded both a knowledge of the science of mechanism and a professional and practical knowledge of the purposes and means of engineering. The need to study mechanical actions in their simplest forms had led to the development of the Engineering Laboratory, in which objects sought were as purely scientific as in Chemical and Physical Laboratories and equally fitted for university teaching and instruction. The advantages which Cambridge had to offer would, in his opinion, ensure the success of a school established on these lines and properly equipped, and whatever the immediate effort required it would be more than justified by the results which could accrue from the

[1] Fly-sheets were written by O. Reynolds, H. Darwin, R. T. Glazebrook, A. R. Forsyth, C. F. Findlay, J. Lyon, and R. F. Scott. They are preserved in the University Archives, Professor of Mechanism and Engineering, No. 53.

connection it would open up between the university and one of the largest and most important of the professions.

The longest of the fly-sheets, and one of the most interesting, was by C. F. Findlay of Trinity Hall, a civil engineer who practised in London and India. He had graduated as tenth wrangler in 1876 and was tutor of his college until 1879.[1]

...The necessity for special scientific knowledge on the part of engineers is largely a growth of the last twenty or thirty years, and becomes daily more imperious. There is no place in the world better fitted to meet that necessity than Cambridge by her associations and the spirit of her studies, and it was inevitable that those who recognise the necessity of such teaching should ask Cambridge to provide it, thus only following at a dignified distance the example of nearly all the other Universities both in the United Kingdom and abroad. That the unfortunate career of the workshops should have surrounded the question with an atmosphere of prejudice and personal hostility is perhaps natural, but it would be most unworthy of the University that a serious educational proposal should be decided on irrelevant issues of that kind.

Apart from petty considerations such as these I should have some difficulty in divining on what grounds Cambridge men, and especially Cambridge mathematicians and men of science, could oppose the reasonable demand for engineering teaching in the University. I am told, however, that the grounds of opposition are two, viz. (1) expense (2) the idea that it is not the proper function of a university to give such teaching.

A well furnished mechanical laboratory is a most useful adjunct for experiment and illustration, and if a thoroughly sound course of teaching were once well established, I have myself no doubt that the University would find friends who would provide the funds for building a laboratory, as other Universities and Colleges have done, without trenching on her present revenues. Cambridge has as many wealthy friends as any University in the world, and there would be no difficulty in finding the money for a laboratory if it were once recognised that the task of providing instruction was being undertaken in a serious and intelligent spirit.

As to the second objection, viz. that Engineering is not a proper subject for a University to meddle with, it seems to me so astonishing a view that I really have some difficulty in treating it seriously. I can only suppose that it is entertained by people who suppose the teaching of engineering to be the teaching of handicrafts, and who have no idea whatever of the kind of knowledge an engineer ought to have to do his work properly.

The training a University can give bears upon an Engineer's work in much the same way that it does upon that of a physician. It might be said with as much

[1] He died in Calcutta of cholera in 1903, at the age of 49.

truth of either the medical, legal or clerical professions that they should not have University teaching provided specially for them as of Engineering. Yet large sums have been spent on such teaching without a protest; and on branches of it which would be useless except for men intending actually to enter those professions.

I could better have understood such objections being raised in Oxford where literary and philosophical studies have until recently held almost exclusive command; but in Cambridge, whose distinction is so largely scientific, it is difficult to believe that they can prevail. Even leaving aside the propriety of special professional training, there are no studies more appropriate to Cambridge than the branches of applied science which Engineering education requires. Nor are there any studies which afford more scope for original research and scientific investigation. New scientific problems are constantly suggested for the developments of Engineering progress. The mutual dependence of pure and applied science on each other is just as close in those branches of science that bear upon Engineering construction as in all other branches. Those properties of materials which are investigated in a mechanical laboratory are just as much a proper study for a University as those investigated in a physical laboratory.

In conclusion, I think I should express the opinion of most Engineers if I say that while of course it is not the business of the University to teach Engineering, which is an art, and not a science, we may fairly ask you to teach those branches of science on which the practice of engineering rests, and which Cambridge is peculiarly adapted to teach well.

Two days before the Grace for extending the powers of the Syndicate came up for approval Scott published his reasons for opposing it. He began by disclaiming any desire to limit the usefulness of the university; on the contrary he wished to see its work developed on sound lines, and his objections to the proposed Engineering School were mainly financial. The studies of the university had recently been greatly extended and it was notorious that the funds available were insufficient to meet the new demands on them. The foundation of the provincial universities and colleges had introduced a new factor into English education and it seemed to him that a division of labour between the old and new establishments was the wisest course. The provincial universities in the industrial centres with their great workshops and eminent engineers ready at hand seemed to him well fitted to give an engineering training, without the waste of machinery and men which would be inevitable at Cambridge. He therefore proposed that the university should, at least for the present, avoid any new departure involving large expenditure and devote its energies and resources to developing the schools it already possessed. To do otherwise would merely add one more starved department with an ill-paid staff to those they already possessed, with the result of retarding general progress.

The last fly-sheet contained two memoranda written at the request of Glazebrook by A. W. B. Kennedy and J. Hopkinson. Kennedy was generally recognized as the leading authority on engineering education at this period, and a brief account of his background seems justifiable. He was born in Stepney an 1847 and educated at the City of London School and the School of Mines in Jermyn Street, where Robert Willis was lecturing on mechanism. At the age of sixteen he was articled to a firm of engineers in Millwall and worked on the construction of twin-screw marine engines designed for blockade runners of the southern States in the American Civil War. Later he worked as a draughtsman for firms in Barrow-in-Furness and Leith and then set up a consulting practice as a marine engineer in Edinburgh. In 1874 he was elected to the chair of Mechanical Engineering in University College London. His appointment was a milestone in the history of engineering teaching. He devised a method of instruction based on experiments, and the term 'engineering laboratory' originated with him. His ideas were quickly accepted by leading engineers throughout the country, and when he retired from his professorship in 1889 laboratories after the London model had been established in Birmingham, Bristol, Sheffield, Leeds, Manchester, Liverpool, Dundee and Edinburgh as well as in universities in Europe, America and Australia. His memorandum was as follows.

From the purely theoretical, or at least unexperimental, side the College can teach the sciences on which Engineering is based, Mathematics, Mechanics, Physics, Geology, and it does much if it teaches these well, and of course with special reference to the particular applications of them which engineers require. But I am sure that a college can do much more than this. It can deal with the experimental side of engineering work. There are a very large number of what may be called technical problems which are in fact merely problems in Physics or Mechanics of a very complicated kind. For instance the whole subject of elasticity as applied to structures and machines is as difficult mathematically as it is important technically. Problems relating to it have to be practically dealt with every day, but scientific experiment has as yet been made only with the very surface of the subject. Problems connected with heat engines are also dealt with every day by the engineer but his data (although more complete than in the case of elasticity) require very great enlargement before they can be anything like complete. In Cambridge there exist, as nowhere else, all the appliances for making men conversant with the higher parts of the mathematical and physical theories involved. No other place that I know of can compare in this respect with Cambridge. At Cambridge therefore of all places it would appear to me important that, if engineering science is taught at all it should be taught experimentally in the sense which I have above indicated. I think the result of the establishment of the necessary engineering laboratories for this purpose in such a place as Cambridge would be really a most important gain to engineering education in England.

John Hopkinson, who wrote the second memorandum, was an even more important figure than Kennedy in the history of engineering at Cambridge, both in his own right and as the father of Bertram Hopkinson, the third Professor of Mechanism, and his career must be described in some detail.

He was the eldest son of a Manchester engineer of the same name who had married the daughter of a prosperous Yorkshire cotton spinner. There were thirteen children by the marriage, of whom five sons and five daughters reached maturity. Three of the sons won distinction in the engineering profession, one was called to the bar and became a Member of Parliament, Principal of Owen's College and Vice-Chancellor of Manchester University, and one achieved a large practice as a doctor. John Hopkinson the elder served his apprenticeship in a firm of millwrights and engineers in which he eventually became a partner, but in 1881 he set up a private consulting practice with his third son Charles. He was a member of the Manchester City Council from 1861 until his death in 1902, and was mayor in 1882. He was unusual among the engineers of his day in recognizing the importance to industry of the education given in the ancient universities, and sent three of his sons to Cambridge and one to Oxford.

John Hopkinson the younger was born on 27 July 1849. He was educated at a boarding school which moved from Cheshire to Hampshire under a headmaster who made a name for himself as a pioneer in the teaching of science to schoolboys. At the age of fifteen and a half, having decided to follow his father's profession, he joined Owen's

College, at that time almost the only place in England outside London which provided a course specifically designed for the education of engineers. The teaching staff included several men of outstanding ability, among whom were two Senior Wranglers, and John soon attracted favourable notice by his industry and intelligence. Early in 1867 he sat for a mathematical scholarship at Trinity College, Cambridge, and came first out of thirty candidates. At about the same time he matriculated at London University, which was then a purely examining and not a teaching body, and while reading mathematics for the Cambridge tripos worked concurrently for the London B.Sc., which he took in 1868, winning the exhibition for Natural Philosophy and Chemistry. Three years later, while still an undergraduate at Trinity, he was admitted to the London degree of Doctor of Science in pure and applied mathematics. In Cambridge he coached with E. J. Routh and under his brilliant tuition swept the board, winning a Sheepshank's Astronomical Exhibition, the Scholarship in Mathematics at London, a Whitworth Foundation Scholarship and emerging in 1871 as Senior Wrangler and first Smith's prize man. Nor did he confine himself to purely intellectual activities; he was a good oarsman and became captain of the Second Trinity Boat Club, and three weeks before the tripos examination won the mile race in the athletic sports, breaking the previous university record for that event. He was not a man who made friends easily, but he became intimate with some of the leading mathematicians of the day, including his contemporary J. W. L. Glaisher and James Stuart, who was six years his senior. The religious tests for fellowships had just been abolished, and he was one of the first Nonconformists to be elected by Trinity under the new statutes.

On leaving Cambridge he started work in his father's factory, but in March 1872, though only twenty-two years of age, he was appointed engineer and manager of the optical works of Chance Brothers, glass makers and lighthouse engineers of Birmingham. A year later he resigned his fellowship at Trinity and married Evelyn Oldenburg, daughter of a naturalized German, with whom he had been in love since he was an undergraduate. The young couple settled in a house near the Chance works at Smethwick and there, in two large rooms which he converted into laboratories, he carried out research into optical problems and conducted experiments in electricity and magnetism for which he was elected a Fellow of the Royal Society in 1878.

In 1877 he moved to London and set up in practice as a consulting engineer. By this time there were three children, the eldest, Bertram, having been born in 1874. Money was short at first, as he had lost nearly all his savings in an iron company of which his father was director, but he soon built up a lucrative practice in the law courts as an expert witness on patents and other scientific matters. He also continued his private researches, and became particularly interested in the theory of dynamos and in power distribution for electric lighting, which was then coming into general use. He became a member of the Institution of Civil Engineers in 1877 and of Electrical Engineers in 1881, and in 1887 was elected to the Athenaeum Club as one of the nine men chosen annually for eminence in literature, art or science. In 1890 he received the Royal Society's medal for his work on the magnetic properties of iron. As his reputation and wealth increased he made many distinguished friends, and in entertaining them, as in all his other activities public and private, he was aided by his wife, a woman of high intelligence and strong personality. He bought a fine house on Wimbledon Common and there his children—three sons and two daughters—grew up in an atmosphere of happiness and well-being. His work brought him into contact with most of the leaders of science and industry and among the frequent visitors to the house were Lord and Lady Kelvin, Sir Benjamin Baker, builder of the Forth Bridge and the Assouan Dam, Sir William Crookes the chemist and many others.

When Stuart announced his intention of retiring the electors to the chair of Mechanism tried to persuade Hopkinson to take his place, but he had too many commitments to be willing to take on a full-time post of this nature. Instead he accepted an offer from London—the Professorship of Electrical Engineering in King's College. This involved no teaching duties and fitted in well with his consulting practice, as he was able to carry out much of his personal research in the Siemens' Laboratory in the college while supervising that of his students. This did not mean, however, that he had lost interest in his old university. He had always held the view that Cambridge ought to play a more active part in the education of engineers, and when the crisis arose over the workshops he willingly accepted an invitation to become a member of the Enquiry Syndicate. After attending its first meeting he put his views on paper in the memorandum attached to Glazebrook's fly-sheet.

A university Engineering School should, he wrote, serve two classes of students—those who intend to become professional engineers and those who value the knowledge for other reasons. In the latter class he mentioned barristers, managers in industry and even clergymen, all of whom benefit greatly from some knowledge of mechanical operations. For them he thought that a workshop run on Stuart's lines was admirable; for the professionals he was less certain.

> The majority of Engineers in the country must be practical men who can deal effectually with the actual execution of a limited class of work, and for these it is probably best that their workshop education should begin earlier than the usual time of leaving the University. But there is also room in the Profession for others who have a wider education addressed to enabling them to deal with circumstances for which there is no direct precedent. For this I believe that the mathematical education of the University would be of great value, but that with it should be coupled constant contact with the physical phenomena in some shape or other, so that the men may realise that every bit of their mathematical knowledge has its counterpart in objective facts...If I were sending a son to the University, who was afterwards to practise the profession of an Engineer, I should not send him to the Mechanical Workshop as at present arranged, but let him work at mathematics and in the Physical Laboratory. What I take it is really wanted is that in addition to the Workshops there should be a Physical Laboratory addressed to dealing with those matters which have a particular interest to engineers, such for example as the investigation of the mechanical properties of materials, the thorough testing of steam and gas engines, hydraulic experiments, the testing from every point of view of dynamo machines, but not their manufacture...

The arguments of Hopkinson and the other protagonists of an Engineering School prevailed, and when the request for extended powers for the Syndicate was submitted to the Senate on 3 June it was approved by 86 votes against 41.

On the following day the Engineering Syndicate reported under their original terms of reference in a special number of the *Reporter*. In an introductory section they reiterated their belief that the continued existence of a School of Engineering in Cambridge and its further development were highly important, but they thought the part which the workshops should play in the training of engineering students and their relation to an engineering laboratory would be best considered after the new professor had been appointed. Certain general principles ought, however, to be determined without delay, and these were dealt with in Part I of the report under the heading 'The Management of the Workshops'.

The Syndicate disliked the division of responsibility between the professor, the superintendent and the Mechanical Workshops Syndicate and recommended that the professor should be directly responsible to the university for all the details of management. They also had grave doubts about the position of the workshops as a trading concern. They realized the need to find employment for the workmen during vacations and when they were not occupied in teaching, but they thought the staff should be reduced to the minimum required for the efficient instruction of students and that work for profit should be limited accordingly. An exception might be made in the case of one or two instrument makers whose services would be of great value to members of the university engaged in scientific research. Under the new regulations which they proposed the responsibilities of the superintendent would be greatly reduced, and they recommended that when Lyon retired (and he had announced his intention of doing so when Stuart left) he should be replaced by a second demonstrator. If these proposals were accepted the Workshop Syndicate might be considered redundant, but if the Senate rejected this view its powers should be increased so as to give it effective control over the trading activities of the department.

The second part of the report dealt with the financial position of the workshops and contained balance sheets and profit and loss accounts for the years 1888 and 1889. The Syndicate complained that the books had been kept in a form which made it impossible to check the items required for the final accounts, and they demanded a complete reform of the book-keeping system. The accounts showed a deficit of just over £125 for 1889, and as trading activities were now to be greatly curtailed there was no likelihood of a surplus in the future, and the Syndicate proposed that the attempt to pay interest on the capital invested in the workshops should be abandoned.

The report's proposals were summarized in seven recommendations which, if approved, would take effect when the new professor was appointed. The Mechanical Workshops Syndicate was to be abolished and the professor was in future to be directly responsible for the details of the management of the workshops. The plant, equipment and stores were to be revalued, but payment of interest on the capital was to be discontinued. The regulations concerning the office of superintendent were to be rescinded when Lyon retired and new regulations drawn up for the practical instruction of the students. The report ended with a complete draft of new regulations for the management of the workshops. It was signed by all members of the Syndicate. An Appendix contained the memorandum by Hopkinson already circulated as a fly-sheet and summarized above and others by Reynolds, Stuart and Horace Darwin.

Osborne Reynolds's memorandum began with a brief description of the Engineering School at Manchester, which had then been in existence for twenty-two years. Until recently instruction had been given only by courses of lectures, mostly on applied mechanics in theory and practice. In 1885, however, an engineering laboratory had been established and experiments were now conducted on problems arising on various branches of the subject. A certain amount of handicraft was carried on, but no work was done for outside bodies and there was no trading. The laboratory testing machines were, however, open to the public and no charge was made for work done on them. About forty students attended during the day and half that number in the evening.

On 22 February Reynolds had been shown round the Cambridge workshops by Lyon and his account of the visit is of particular interest as that of an unprejudiced observer who had not been involved in the recent quarrels.

> It seemed to me that these Workshops are most excellent for the purpose of giving instruction in handicraft. I think they are managed on an admirable plan; and that everything seems to be done in a very satisfactory way. As far as I can judge, they are working well, and, I cannot but think, must be very advantageous to the student. The system of instruction is different from ours at Manchester, as our principal object is the Laboratory, to which handicraft teaching is with us only secondary.
>
> But the system at Cambridge seems to me to be very good; and the large staff of skilled workmen for the instruction of the students affords them an admirable

opportunity; a better one even than they would get on going into business works; indeed pupils on going into works are left to pick up what they can; and those with no special mechanical aptitude often spend a year or two with but very little profit. I am surprised that such efficient instruction can be given at so little cost.

Nevertheless, he thought the Cambridge system should be modified, and that theoretical and laboratory teaching should be increased. An Engineering School which brought able mathematicians into the profession might become an institution of national importance, though he did not think it would ever attract large numbers.

The last two memoranda can be treated more briefly. Stuart's was a reply to those of Hopkinson and Reynolds. He thought they both underestimated the value of workshop training, but agreed that a laboratory was also most desirable. He claimed that it was only lack of funds which had prevented him from establishing one as early as 1880. Darwin made a strong plea for the transformation of the Mechanical Workshops into an Engineering Laboratory, but it is unnecessary to repeat his arguments as most of them had already been put forward by Kennedy and Hopkinson.[1]

The publication of this report brought to an end the long series of disputes and arguments about the future of the Department of Mechanism, and when it came up for discussion on 7 June the only speech which need be referred to was by Lyon, who formally announced his intention of resigning from the post of superintendent when Stuart left, saying, 'The *imperium in imperio* which had existed for the last five years or so could only work under the circumstances for which it was created.'

During the Long Vacation tempers cooled, and when the seven recommendations were put to the Senate on 16 October they were approved without opposition. The way was now clear for Stuart's successor, and the path which he was expected to follow had been clearly marked out for him.

There had at one time been some doubt as to whether Stuart would have a successor, as the wording of the relevant statute was far from clear. The Council felt it necessary to refer the matter to the Chancellor,

[1] He quoted largely from a paper on Engineering Laboratories which Kennedy had read to the Institution of Civil Engineers and from the discussion which followed. See *Proc. Inst. Civ. Engrs*, LXXXVIII (1887), 1.

the Duke of Devonshire, who, on 17 June 1890, gave his formal decision that the Professorship of Mechanism was in fact 'part of the Permanent Professorial Staff of the University'. When this decision was made known the electors sounded Hopkinson about his willingness to accept the post, but as already stated he refused to stand. When asked to suggest an alternative candidate he replied, 'Try Ewing of Dundee; I can think of no one better, or so good.'

Candidates for the vacant chair were asked to submit their names by 8 November, and when the list closed there were thirteen candidates.[1] The Electors were the Vice-Chancellor, W. Airy, Dr Besant, Sir F. J. Bramwell, Professor Cayley, Mr Martin, Dr Phear and Lord Rayleigh. No record of the voting has survived, but Professor Cayley in a letter to the Registrary said that the result was reached by a process of successive elimination. The professor elected was James Alfred Ewing of Dundee.

Stuart's political career after leaving Cambridge was honourable but not particularly distinguished. According to his friend G. F. Browne, who was a staunch Tory, he fell into the hands of the extreme left wing of the Liberal party and lost the confidence of its leaders, though at one time Gladstone, with whom he was on terms of some intimacy, considered him seriously for a post which might have led to very high office, He had a great capacity for making friends and led a full, active and happy life. He was Lord Rector of St Andrew's University from 1892 to 1901 and in 1909 his long years of public service were recognized by his appointment to the Privy Council.

Soon after his resignation from the professorship he married Laura Elizabeth Colman, described by Browne as 'one of the best of our early Newnham students'. On the death of his father-in-law in 1898 he became a director of the mustard firm of J. and J. Colman, Ltd., and moved to Carrow Abbey in Norwich. He died there on 12 October 1913.

[1] Wilfred Airy of London; Professor G. F. Armstrong of Edinburgh University; T. H. Blakesley of King's College; Professor J. A. Ewing of University College, Dundee; M. E. Fitzgerald of Queens' College, Belfast; Professor W. Garnet of Durham College of Science; R. T. Glazebrook of Trinity; Professor T. A. Hearson of the Royal Indian Engineering College, Coopers' Hill; James Lyon of Clare; R. E. Middleton of London; W. N. Shaw of Emmanuel; Rev. F. Smith of Trinity College, Oxford; Professor R. H. Smith, of the Mason Science College, Birmingham.

The departmental workshop still performs most of the functions for which Stuart created it. It makes a large proportion of the equipment used for teaching and research in the department of Engineering and a considerable amount for other university departments and for research establishments elsewhere. Classes are still held in it for selected undergraduates, though most of them now obtain their workshop qualifications in industry during the Long Vacation. Nevertheless, the workshop probably constitutes Stuart's principal memorial in the Engineering School of which he was the first professor.

4 JAMES ALFRED EWING

Professor of Mechanism and Applied Mechanics, 1890–1903

THE YEARS BEFORE CAMBRIDGE, 1855–1890

James Alfred Ewing was born on 27 March 1855, the third son of James Ewing, Minister of St Andrew's Church in the Cowgate, Dundee. James senior (the son was known as Alfred) entered Glasgow University at the early age of fourteen, and after completing courses in arts and divinity was ordained into the presbyterian ministry in 1833. Four years later he was appointed to the Cowgate Church, and in 1847 married Marjorie Ferguson, daughter of a Glasgow solicitor. Five of their children reached maturity, and Marjorie, who had enjoyed a more liberal education than was usual for girls in those days, gave them a good grounding before they went to school and inspired them with a genuine love of learning. All the boys won scholarships to a university.

In a family whose chief interests were religious and literary the young Alfred found his pleasure mainly in machinery and experiments, and at the age of twelve announced his intention of becoming an engineer. Little science was taught at the Dundee High School which he attended, but he spent his pocket money on tools and chemicals and converted an attic in his home into a laboratory. 'It was in this irregular fashion', he wrote in later life, 'that I began to explore the pleasant borderland of physics and engineering where I have roamed happily for many years.' But his interests were not confined to technical subjects; he was a voracious, though discriminating, reader, and his love of good literature contributed to the making of a first-rate conversationalist and an easy and attractive writer. His hard work at home and at school were rewarded when, at the age of sixteen, he won a scholarship to Edinburgh University.

Undergraduate teaching in the Scottish universities followed the continental system of professorial courses on individual subjects in contrast with the general syllabus directed by tutors favoured by the

Oxford and Cambridge Colleges. These professorial courses were highly esteemed, but during the eighteenth century the importance attached to the degree itself declined, and until late in the nineteenth century only a small proportion of the students applied for graduation.

A Professorship of Engineering had been established at Edinburgh in 1868, only three years before Ewing matriculated. The first holder of the chair was Fleeming Jenkin, a professional engineer who specialized in the development of electric cables for marine telegraphy. He was a friend of Sir William Thomson, later Lord Kelvin, and for three years before coming to Edinburgh had been Professor of Engineering at University College London. His lectures constituted a course in the Science Faculty, but to qualify for the degree of Bachelor of Science (Engineering) candidates had also to show proof of a liberal education by producing qualifications earned elsewhere or by obtaining two certificates in the Faculty of Arts. The degree examination was in two parts, the first in General Knowledge in Mathematics, Natural Philosophy and Chemistry, and the second, taken after an interval of six months, in Mathematics applied to Mechanism, Engineering and Drawing. An increasing number of students were now taking these examinations and proceeding to the degree.

Ewing chose Engineering as his principal subject, but he also studied Natural Philosophy under Professor Tait and English Literature under David Masson, a friend of Thomas Carlyle. At the end of his first session he received a letter from Jenkin asking if he would be willing to check the answers to questions which he set each week to his students. This offer was made in view of an exceptionally favourable report from Professor Tait, and it marked the beginning of a close association between teacher and pupil. Ewing became a frequent visitor to the Jenkins' house and acquired a lasting affection for both the Professor and Mrs Jenkin.

The Scottish universities confined their teaching to the six winter months so that the poorer students could earn money to pay their fees by taking employment during the summer. The professors used the long vacation in various ways, some, as Ewing described it, to enjoy 'an agreeably protracted period of undisturbed contemplation', others, of whom Jenkin was one, in business or professional activities.

Jenkin was now in partnership with Sir William Thomson as engineer to the Great Western Telegraph Company, and at his suggestion Ewing obtained work during the summer of 1872 in the cable factory of a subsidiary company in the Isle of Dogs near Greenwich. He travelled south by freighter and was placed on the pay roll at a salary of a pound a week. He returned to Edinburgh in the autumn for his second session but resumed work in the factory during the following summer. In February 1874 he received a sudden invitation from the Telegraph Company to supervise the repair and relaying of the inshore ends of some of their cables off the coasts of Brazil and Uruguay. He carried out this task with conspicuous success, and on his return the recommendations of his report were accepted by Sir William Thomson in their entirety. During the summer of 1874 he continued work on the cable problem in the Siemens factory at New Charleton and in October returned to Uruguay to supervise the laying of the cables on the lines he had suggested. In spite of severe gales and a local revolution he completed his task in August 1875 and then returned to Edinburgh to resume his interrupted studies.

During the next three years he followed the normal courses in the university and also helped Jenkin in various research projects, including one on gas engines. Much of his work was done in Jenkin's house, and among the people he met there was Robert Louis Stevenson, a close friend of the family. He also continued his association with Sir William Thomson, who was then President of the Royal Society of Edinburgh, and occasionally visited his laboratory and that of his brother James, the Regius Professor of Engineering in Glasgow University.[1] In the winter of 1876 he gave a course of lectures on the elements of engineering at the Watt Institute in Edinburgh to an audience of about thirty men from local offices and workshops. He found that the art of lecturing came easily to him and decided to make the teaching of engineering his profession. He applied unsuccessfully for the chair of Applied Mechanics and Civil Engineering at McGill University in Montreal when it fell vacant in 1877, but in the following summer he

[1] The Regius Professorship of Engineering at Glasgow was established in 1840. It was the first university Professorship in Engineering to be founded in Britain, though antedated by a few weeks by the Professorship of the Arts of Construction in connection with Civil Engineering and Architecture which was established by King's College, London, on 10 July 1840.

was more fortunate. At a dinner party given by the Jenkins he met a Japanese official who had come to Scotland in search of a Professor of Mechanical Engineering for the new University of Tokyo. He applied for the post and was accepted for an initial period of two years. Before leaving Scotland he was admitted to the degree of B.Sc. (Engineering), seven years after he had matriculated.

When Ewing reached Japan only a decade had passed since the revolution of 1868 put an end to feudalism and two centuries of isolation, and less than a year since the suppression of the last rebellion of the supporters of the old regime in the province of Satsuma. During this time the process of westernization had been swift and ruthless, and Tokyo University was playing an important part in the re-education of the new governing class. It was situated in the suburb of Hitotubasi and housed in two-storied buildings enclosed by a palisade. There were two hundred students, most of them members of the warrior clan of Satsuma, and the main faculties were law, literature and science.

Ewing's class originally numbered sixteen, and he lectured for twenty hours a week on the elements of mechanical engineering and thermodynamics. The presentation of new ideas in a language foreign to the students was no easy task, but by the end of the first week he felt that he had got the situation under control. His optimism was justified, and when he left the country five years later he had laid the foundations of a school of engineering which was to occupy a key position in Japanese technological education. Some of the best of his students embarked on research under his supervision and a few of them eventually became engineers with an international reputation. Ewing himself carried out research on the earthquakes from which Japan suffers so frequently and devised a new type of seismograph. A paper which he wrote on this subject was read to the Royal Society by Sir William Thomson. Later he studied phenomena connected with magnetism to which he gave the name of hysteresis, and again Thomson presented the paper to the Royal Society. In 1881, owing to the illness of the Professor of Physics, he took over the teaching of electricity and magnetism to advanced students. At the same time he renewed his contract with the university for a further period of two years.

In May 1879 he had married the stepdaughter of the American

Professor of Moral Philosophy, Annie Washington, a direct descendant of the first President of the United States. The birth of two children in the early years of their marriage seriously affected her health, and this was one reason which led him to refuse the offer of a further extension to his professorship. He felt, too, that he was in an academic backwater, and that his personal research was being handicapped by the limited facilities available in Tokyo. However, the event which finally sealed his determination to return home was the establishment of a Professorship of Engineering in the new University College of Dundee. He obtained the support of Sir William Thomson and Fleeming Jenkin for his candidature, and on 4 November 1882 received a telegram announcing his election. The Japanese authorities had agreed to release him before the expiration of his contract in the event of his success, and when the news came through they presented him with a glowing testimonial for his services. He returned to Scotland via the United States, visiting on the way his wife's parents, who had returned to Philadelphia, and her family's estate in West Virginia, now sadly impoverished as a result of the Civil War.

He reached Dundee in time for the opening of the College by Lord Dalhousie in October 1883. His work was at first confined to lectures on civil and mechanical engineering, but later he added lectures and practical work on electrical engineering. There is no suggestion that he did not take his professorial duties seriously, but he was ambitious, and it is clear that he regarded his post in the small provincial college as a stepping stone to higher things. Outside the college he devoted much attention to the work of the Dundee and District Sanitary Association, which sponsored an agitation for proper sanitary arrangements in the working-class areas of the town where conditions provided a humiliating contrast to the clean and airy houses of the workers in Tokyo. A sharp earthquake in the eastern counties of England revived his interest in seismography, and he was invited to carry out a series of tests with his own instruments (now manufactured for him by the Cambridge Instrument Company) in an observatory on Ben Nevis. He also used his seismograph to measure the movement of the new Tay Bridge during the passage of trains, which opened up a new field of study for him.

Fleeming Jenkin died suddenly in June 1885, and Ewing, as one of the trustees, helped to edit his literary and scientific papers. About this time he was invited to write an article for the *Encyclopaedia Britannica* on the strength of materials, and when submitting his draft he suggested the need for one on steam and other heat engines. His offer to write this was accepted, and the article later formed the basis of his most successful textbook. He also continued his work on hysteresis, and in May 1887, on the strength of his various publications, was elected a Fellow of the Royal Society. This brought him into contact with many scientists and engineers in England, including John Hopkinson and Oliver Lodge. He also became an active member of the British Association, and at a meeting in Leeds in 1890 his paper on the molecular theory of induced magnetism and an ingenious model to illustrate it attracted much attention. In the meantime he had added considerably to his income by building up an extensive practice as a consulting engineer.

In August 1889 James Thomson resigned the Regius Professorship of Engineering at Glasgow University owing to failing eye-sight. Ewing applied for the post, but to his great disappointment was not successful. However, in the following June he learnt of Stuart's impending resignation from the Professorship of Mechanism at Cambridge, and decided to apply for it. His candidature was supported by Sir William Thomson, J. J. Thomson of the Cavendish, John Hopkinson, Osborne Reynolds and other eminent members of the profession. On 12 November 1890 he received a telegram from the Vice-Chancellor saying, 'You are elected Professor.'

THE ENGINEERING LABORATORY AND THE TRIPOS, 1891–1894

The Michaelmas Term was more than half over when Ewing received the news of his election. He was not required to take up his duties until after the Christmas vacation, but he came south as soon as possible to study the situation and find a house for his family. He was well aware of the controversies which had led to Stuart's resignation and knew that he would have to proceed with tact and caution if he were to succeed where Stuart had failed. He was fortunate in being invited by Dr Porter to be his guest in the Lodge in Peterhouse while making his preliminary survey, and he remained there until he was able to move into a house in Brooklands Avenue. His wife and children joined him in the following May. Porter possessed great influence in the university and his experience on the Workshops Enquiry Syndicate was of immense value to Ewing during these months of initiation.

Ewing decided to put his aspirations and plans to the university in the form of an Introductory Lecture, and he delivered this under the title 'The University Training of Engineers' on 20 January 1891. He had been a Professor of Engineering for twelve years before coming to Cambridge, and he brought with him the Scottish tradition of professorial teaching, Kennedy's idea of the engineering laboratory, and his own experience as practical engineer, teacher, research worker and academic administrator.

After paying tribute to his predecessor in the chair of Mechanism he went straight to the heart of his discourse. 'I have to plead for nothing less than the inclusion of a complete School of Engineering in that new Cambridge which is fast springing up within the old.' He compared the educational requirements of engineers with those of doctors. Both needed a threefold system of training—first a basis of general scientific study, then a middle structure of specialized science applied to professional practice and lastly the acquisition of the practical knowledge which can only be gained by experience. 'The chief aim of a University School of Engineering is to give such a training in pure and applied science that when the pupil follows it up by spending two or three years in contact with actual constructive work he is well fitted to begin his professional career.' He compared the English system of practical instruction in industry, with absolutely no place for theoretical instruction, and the continental method, in which theoretical training bulked so largely as to leave little or no room for practice. He gave a brief description of the evolution of the Polytechnics and Hochschulen on the continent and compared their methods with recent developments in England. Fifteen years previously the teaching of engineering in England was confined mainly to the lecture-rooms; now, thanks to Professor Kennedy, no school was considered complete without a costly laboratory. 'Kennedy found the engineering laboratory no more than a means of private research. He left it an instrument of education.' It had three objects—first, to give the student practice in experimental work which he might use later, secondly to teach him how such work is conducted, and thirdly it contributed to the stock of common knowledge by furnishing opportunities and appliances for original research. 'Put a good student into a laboratory and you inspire him with the enthusiasm of the investigator; put a good teacher into a laboratory and you will ensure that he will never cease to be a student.'

Laboratory work, said Ewing, was essentially measurement, and a fully equipped engineering laboratory required costly equipment. He was prepared to start on a small scale, and the existing workshops would provide a valuable nucleus, but he hoped he would be acquitted of any intention to rest content until a laboratory was provided worthy of the University of Cambridge. On the vexed question of workshop training expert opinion was divided, but the general drift seemed to him to be in favour of a limited amount of practical instruction, and he proposed to make use of the existing shops by continuing a series of exercises in wood and metal working. All this, however, was largely a matter of detail—the important thing was to get the engineering school accepted as an integral part of the university system.

The work done in my department must be closely interlinked with that done in other departments; the teaching of applied mechanics must go hand-in-hand with the teaching of mathematics, physics, chemistry; the students of engineering must receive such academic encouragement as will attract the best men...We must have a department working in the closest association with collegiate tuition and other spheres of university activity, amply furnished in respect of buildings and plant, secure of University favour, and forming an accepted avenue to University distinction.

He ended with the cry of 'Give, Give' not only to those whose business it was to sit upon the lid of the University Chest, but to all friends of Cambridge, to all friends of engineering education. He appealed for that help earnestly and confidently, convinced that a laboratory was not simply desirable, but indispensable.

A report by the Workshops Enquiry Syndicate published on 9 February contained a detailed statement of the equipment needed to enable laboratory teaching to be started as soon as temporary accommodation could be provided. Ewing wanted to set up a laboratory of applied electricity in addition to one of mechanical engineering and asked for £250 for electrical apparatus as well as money for the purchase of a small steam engine. He also suggested that when Lyon retired on Lady Day the post of Superintendent of the Workshops should be abolished and replaced by a second demonstratorship. The Syndicate supported these requests and they were approved by the Senate on 12 March.

It was fortunate for Ewing that at this time, when the department needed room for expansion, the ideal site became available. In March 1886 the governors of the Perse School for Boys, which lay along Free School Lane immediately to the south of the Mechanical Workshops, offered the freehold to the university for the sum of £12,000. The area of the site was 16,767 square feet and the buildings consisted of the original seventeenth-century hall, a block of class rooms built in 1842 and the houses of the head and second masters, all in excellent condition. The proposal to purchase was put to the Senate in February 1888. It was opposed by many speakers in the discussion on the grounds that the university could not afford the money at a time when other land recently purchased still awaited development owing to lack of funds, but J. W. Clark, who was now Secretary of the Museums and Lecture Rooms Syndicate, spoke strongly in favour, and the report was accepted on 3 May by 99 votes against 60.

On 18 December 1890 a syndicate was appointed with Clark as secretary to consider the appropriation to university purposes of the school buildings and also the arrangements for those parts of the adjacent New Museums site which had not yet been allocated. A letter was sent to the Special Boards which were known to be in need of more accommodation and to the Professors of Botany, Experimental Physics and Mechanism asking whether their requirements could be satisfied by the assignment, either temporary or permanent, of space in the Perse School buildings or on the adjacent ground. Ewing wrote a long letter which, together with the other replies, was printed as an appendix to the Syndicate's report. He said

> I find that the buildings in which the Department is at present housed are wholly given over to workshop use. There is a small and not very convenient drawing office which is already over-crowded by the class which the Demonstrator, M. Nicholson,[1] has begun this term to teach in that subject. There is no lecture-room, no laboratory, no museum. A single room serves for office, store and

[1] Appointed by Stuart shortly before his resignation.

Professor's room. Even as workshops the present buildings are poor. They are dark and cramped. They have every appearance of having been designed to be merely temporary. No re-arrangement of the existing rooms can provide the accommodation that is needed for teaching.

It is now recognised that an essential part of an engineering school is an engineering laboratory. The engineering laboratory needs to be large. It must hold a steam engine and boiler specially arranged for experimental work; tanks and other apparatus for hydraulic measurements; apparatus to experiment on the transmission of power; and a number of other bulky appliances. It should include rooms for dynamos, and for experimental work in applied electricity.

The Willis collection of Mechanical Models is at present stored in certain old houses in Free School Lane. Inspection of the models in their present quarters is impossible. The collection has much historic as well as educational interest. It should be transferred to rooms where the models can be seen and properly cared for. It would form a valuable nucleus of an engineering museum...

He added a detailed statement of the accommodation which he needed. His list included two laboratories, each of 3,000 square feet floor area, as well as one for applied electricity, a lecture-room, drawing office and museum, staff and research rooms, a wood-working shop, a fitting and turning shop for metal work, an instrument shop, stores, 'and possibly a small foundry'.

He concluded his letter by stressing the desirability of keeping the engineering school close to the Cavendish and Chemical Laboratories, as much of the work of his students would be done in those places, and the proximity of the buildings would greatly facilitate the arrangement of lectures, etc. He felt sure that a public appeal for funds would be successful and that the money raised would enable the department to make good use of the whole area covered by the school and the garden to the south of it. The extent to which the old school buildings would be retained would depend on the funds available, but if money were short many of them could be adapted by internal alterations and minor additions. The immediate need of the department was for temporary provision for lectures, and housing for the Willis models and laboratory work.

The Syndicate agreed with all Ewing's proposals and recommended that the greater part of the Perse School site should be allocated to the Department of Mechanism, though the headmaster's house was to be assigned to the Department of Botany, and three class rooms were to be used temporarily by the Special Boards for Classics and Medieval and Modern Languages. The main block was to be adapted for Ewing's use. The large school-room would become an engineering laboratory, the smaller school-room adjoining it a drawing office; the class rooms on the ground floor would provide a lecture-room and a laboratory of applied electricity and the second master's house could be used for private rooms for the professor and for a room to contain the Willis models. Later on, if permanent buildings were erected, the greater part, if not the whole, of the area covered by Stuart's workshops could be evacuated.

In the discussion on 28 February the complaint was made that the Syndicate's plan amounted to 'Science first and the rest nowhere', but Ewing defended his claim to the lion's share of the new territory with tact and moderation and was supported by J. J. Thomson and Glazebrook. When the report came before the Senate on 12 March it was approved without a division.

Two days later the Workshops Enquiry Syndicate published a memorandum on the proposed Engineering Laboratory in which they made the case 'for giving the subject of engineering a place in the studies of the University similar to the place it holds in other Universities and Colleges, and for providing the Professor of Mechanism and Applied Mechanics with those appliances which are now recognised as essential to its satisfactory treatment.' The Perse School site had cost the university £12,000 and to build and equip the Engineering Laboratories would cost about £20,000 more. This money could not be found by the university, but

> It may be claimed that the organization of a School of Engineering at Cambridge is a matter of more than local importance. It will enable the exceptional facilities which Cambridge offers for the study of mathematical and physical science to be taken advantage of by students of engineering to obtain a professional and scientific training of the most appropriate kind without sacrificing the general advantages of University residence. It will probably draw not a few men of mathematical ability into a profession they would otherwise not enter. Finally, it may become a training ground for teachers of Applied Mechanics, whose influence in technical schools and classes throughout the country has a wide influence on the industrial population.

The memorandum was signed by the Vice-Chancellor, Montague Butler, and all members of the Syndicate.[1]

It was followed on 11 November by a further report in three parts. Part I contained new regulations for the workshops, which were in future to occupy a subordinate position and would not be used for trading purposes. Part II informed the Senate that a committee was being formed to launch an appeal for funds, and recommended the appointment of an 'Engineering Laboratory Syndicate' to arrange for the further development of the Engineering School in the university. Part III proposed the appointment of another syndicate to consider the establishment of a tripos. The report was approved without opposition on 17 December.

Meanwhile the Perse Buildings had been taken over, and during the Long Vacation of 1891 William Sindall the builder converted the two rooms allocated as temporary laboratories and they came into use in the Michaelmas Term. Much apparatus was made for them in the workshops and gifts of equipment and books were received from Sir William Thomson, Horace Darwin, J. W. Clark, the Institutions of Civil and Mechanical Engineers and the University Press. An appeal Fund for the permanent laboratory was opened by gifts of £1000 from the Chancellor, the Duke of Devonshire, and £500 from Lord Derby, but after this promising start money came in slowly, and when the Engineering Laboratory Syndicate issued its first report in May 1893 only £4,847. 16*s.* 0*d.* had been subscribed. However, this was considered sufficient to justify a start being made, and plans for the redevelopment of the Perse site were drawn up by the architects, Marshall and Vickers.

The Laboratory Syndicate issued its second report on 24 April 1893. The subscription list had increased by less than £100 since March and the money available was only sufficient to cover the cost of the central block and the Steam Laboratory, both of which were by that time well on the way to completion. Certain unnamed members of the Syndicate were so convinced of the urgency of the need for the south wing and the waste of money which would be entailed by providing temporary

[1] J. Porter, C. Cayley, J. J. Thomson, J. A. Ewing, G. B. Atkinson, O. Reynolds, J. Hopkinson, H. Darwin, R. T. Glazebrook, W. N. Shaw, H. F. Newall.

accommodation that they offered to lend, at 3 per cent interest, a sum sufficient to cover the cost of its erection. The Financial Board, however, did not like the idea of a special loan and suggested that the money should be guaranteed as a charge against the building fund of the following year and this was approved without opposition. The south wing was completed before the end of the Easter Term, 1894.

On 15 May the new laboratories were formally opened by Sir William Thomson, now Lord Kelvin. Among the many guests was Professor Kennedy, who received a special welcome from the Vice-Chancellor as the founder of the first engineering laboratory, the model for all subsequent laboratories, including the present one. Lord Kelvin spoke at some length about the equipment which had been provided, and suggested that although a good start had been made bigger testing machines were needed and there was room for further development in the field of hydraulics. Kennedy and other speakers followed and then Ewing rose to reply. He stressed the debt which the new laboratory owed to the workshops. When he came to Cambridge he had had some doubt as to their precise function and utility, but he had no doubts now. Not only were they invaluable for practical work, but he believed them to fulfil an important educational role and he wished very much that Professor Stuart could have been with them now. These remarks were greated with loud applause. He concluded with an appeal for a further thousand pounds to enable them to replace the 'unsightly and unsuitable sheds' which housed the workshops. The visitors were then entertained to tea by Ewing and his wife, after which the laboratories were opened for inspection.

While the new laboratories were being built Ewing had achieved the second great objective which he had put to the university in his inaugural lecture. He had asked that the engineering course should 'be given the freedom allowed to the specialists of the Triposes and should not be denied some claim to a degree', and this was supported by the Workshops Enquiry Syndicate in the report of 11 November 1891 already quoted. In the third section of that report they stated their conviction that

> the arrangements for the education of the students attending the Mechanical Department cannot be regarded as satisfactory or complete until provision is

made for an Honours Examination in the subjects taught in the Department... Some years ago the Senate gave a general approval to a scheme for establishing an Honours Examination for students of Engineering, but up to the present time the details of such an examination have not been approved. The Syndicate are not empowered to report upon such a scheme, but they are unanimously of opinion that the establishment of an Honours Examination specially adapted to the scientific training of students in the Mechanical Department is of urgent importance to the proper development of the study of the Mechanical Sciences in the University. They therefore recommended:

'That a Syndicate be appointed to take into consideration the establishment of an Honours Examination in Mechanical Sciences.'

The recommendation was approved and the Syndicate appointed on 11 December.

In the meantime teaching continued on the lines laid down by Stuart. In the Lent Term 1891 Ewing lectured on the strength of materials and theory of structures and gave laboratory demonstrations on applied mechanics; Nicholson, the demonstrator, taught geometrical and mechanical drawing, and practical instruction on mechanism was given in the workshops. Lyon, as already stated, resigned in March, and Nicholson left in July to become Professor of Mechanical Engineering at McGill University, Montreal, the post for which Ewing had applied unsuccessfully fourteen years earlier. Two new demonstrators, both of whom held the degree of B.Sc. in London University, were appointed on 30 September—W. E. Dalby, who came from the railway works at Crewe, and C. G. Lamb, whose subject was electrical engineering.

The syndicate appointed to consider the establishment of a tripos reported on 23 May 1892. In a short preamble the members stated their belief that

the wants of students of the Mechanical Sciences are not met by any of the existing Triposes. They consider further that until the subjects of its teaching are embraced in an Honours Examination the Department of Mechanism and Applied Mechanics must have its usefulness much restricted, and they are aware that the want of such an examination deters men from coming to Cambridge who would otherwise be likely to profit by the training in science which the University offers.

They proposed the immediate establishment of a tripos, and submitted a complete set of regulations and schedules together with the consequential alterations to Ordinances. The examination was to be in two parts. Part I was comparatively elementary and all the mathematics required was covered by the first paper. Six written papers were to be followed by oral and practical examinations, and success would qualify for the B.A. degree with honours. Part II was more advanced, but candidates would not be expected to take more than two of the four written papers and three practical examinations.[1] This report was approved on 10 November 1892 and the first examination in both parts was held in the Easter Term 1894.

In June 1893 the first inter-collegiate examination in the subjects for the tripos—the forerunner of the modern 'Preliminaries'—was held in the departmental lecture-room. Three second- and third-year men and nine freshmen were candidates. Seven men sat for the first tripos examination in the following year and all obtained honours. There were no candidates for Part II until 1897.

GROWTH AND CONSOLIDATION, 1894–1903

The organization of the laboratories and the establishment of the tripos formed Ewing's main preoccupation during his first years in office, but he still found time to do a considerable amount of private research and consulting work. The first edition of his famous textbook, *The Steam Engine and other Heat Engines*, was published in 1894 and met with immediate success. He developed his theories on hysteresis and was awarded a Royal Medal by the Royal Society for his work on magnetic induction. His seismograph was employed to study vibrations caused by the passage of railway trains, and important results were achieved. He formed a close association with Charles Parsons in developing the steam turbine, and in 1897 took part in the trials of the experimental vessel *Turbinia* when she achieved the unprecedented speed of thirty-

[1] Part I. 6 Papers: Mathematics, Mechanics, Strength of Materials and Theory of Structures, Principles of Mechanism, Heat and Heat Engines, Electricity and Magnetism. Oral and practical examinations in Mechanics and Elasticity, Heat and Heat Engines, Mechanism, Surveying, Drawing, Electricity and Magnetism. Candidates could take all subjects but would not be expected to take more than five written papers and five orals.

Part II. Theory of Structures, Heat Engines and Dynamics of Machines, Hydraulics and Geodesy, Electricity and Magnetism—two papers on each subject. Oral and practical examinations in Engineering Laboratory Work, Geometrical Drawing and Graphic Statics, Machine Drawing and Design, Applied Electricity. Candidates could take all subjects but would not be expected to take more than two written papers and three practicals.

five knots over the measured mile. In the same year he was elected to a professorial fellowship in King's College.

The coming into force of the tripos ushered in a period of departmental expansion which continued with increasing momentum until Ewing left Cambridge in 1903. In the Michaelmas Term 1894 74 undergraduates were reading engineering. Numbers exceeded one hundred for the first time in 1897 and in Ewing's last term as professor the figure was 226. During this period the total undergraduate population in the University declined slightly from about 2,800 in 1894 to about 2,710 in 1903. In ten years the proportion reading engineering thus rose from 2·64 to 8·34 per cent.

The two demonstrators, Dalby and Lamb, proved to be first-rate men, but although they took over much of the responsibility for organizing and equipping the mechanical and electrical laboratories additional staff soon became necessary to cope with the teaching. Two assistant demonstrators were taken on in 1894 and by 1903 the number had risen to eleven. Among them were C. E. Inglis, a future head of the department, J. W. Landon, who became Secretary of the Faculty Board, F. J. Dykes, T. Peel, A. Chapple and A. H. Peake, lecturers and supervisors to many generations of undergraduates.

On 5 February 1894 the Council of the Senate issued the first of a series of reports on post-graduate study which were to have a profound effect on research activities in all departments of the university. The feeling had been gaining ground for some time that an effort ought to be made to attract graduates of other universities to Cambridge by offering courses of advanced study and research, and it was generally agreed that men who completed the courses satisfactorily should be rewarded either by admission to a degree or by the award of a certificate or diploma. A syndicate was appointed to consider the problem in detail, and their third amended report, approved on 3 June 1896, established a new system of post-graduate work and awards in the university. With appropriate safeguards graduates of other universities were admitted either to Courses of Advanced Study or Courses of Research. Advanced Students in the first category (forerunners of the present Affiliated Students) were allowed to sit for a tripos examination after six terms' residence and could then be admitted to the B.A. degree.

Advanced Students who had completed a year's course of research (now known as Research Students) and had written up their results in the form of a dissertation could receive a Certificate of Research, and after six terms' residence they too could be awarded the B.A. degree. Cambridge graduates could take the Research Courses by permission of the Degree Committee of their Special Board of Studies. Six men began research in the Engineering Department under the new regulations in 1897 and the first Advanced Student to obtain a degree in engineering qualified in Part II of the Mechanical Sciences Tripos in June 1899.

The increase in the number of students, the extension of research activities and the duplication of courses for the honours and ordinary degrees soon made the new buildings inadequate for the needs of the department. Only one lecture-room was available for six lecture courses, and it was too small for the larger classes and unsuitable for lectures requiring illustrations. When the Engineering Laboratory Syndicate issued its final report on 9 November 1895 it stressed the urgent need for further accommodation, but nothing effective could be done until more money became available.

In 1896 two important plots of land were purchased by the university —a strip along the northern edge of the New Museums site from a private owner and two acres to the south of Downing Street from Downing College. Four years later six more acres were brought from the college to form what is now known as the Downing site. These deals were not approved without opposition, as some members of the Senate felt that the university ought not to acquire more land while it lacked the resources to develop that which it already possessed. It seems strange today to read that J. W. Clark considered it 'midsummer madness to call the site of Downing College central. It lies on the very edge of the University and has no other advantage to outweigh the defects of its situation. It has no frontage, and the only approach to it is along a narrow and dirty lane which has not even the merit of being straight.' Ewing, it is pleasant to record, spoke in favour of the purchase, saying that even if there were no immediate use for the site no one could foretell what would be wanted in thirty or forty years time.

As soon as the purchase of the first two plots was approved a syndicate was appointed to consider the needs of the scientific departments and prepare a development plan for the whole area. Ewing put in a bid for more land between his existing laboratory and the Cavendish, and his request was endorsed by the syndicate. They proposed to move the Department of Botany on to the Downing site, and when this was done to give the old headmaster's house, the annexe to the herbarium and the garden space between them to Engineering. These and their other proposals for redevelopment were approved on 25 November 1897. The acquisition of this new land and the growing needs of the scientific departments for more accommodation and larger staffs made it essential for the university to find new funds for capital development. In February 1897 the university and colleges launched a public appeal, and the queen agreed to act as patron of the sponsoring body, which was named the Cambridge University Association. By the end of the year contributions to the Benefaction Fund amounted to just over £50,000. Part of this sum would in due course have been allocated to Ewing for his north wing, but in the event he obtained the money in an altogether different, unexpected and tragic manner.

In August 1898 he took his wife and children to Switzerland for a mountaineering holiday with John Hopkinson and his family. All five Hopkinson brothers were members of the Alpine Club, and under John's expert tuition Ewing was initiated into the art of rock climbing. On 27 August Hopkinson set out with his son Jack and two of his three daughters to climb the Petite Dent de Vesivi, a rock which apparently offered little difficulty to experienced climbers. Ewing, who was feeling stiff after his first real climb on the previous day, decided not to go with them. When the party failed to return search parties were sent out, but it was not until the following morning that the four bodies were found, roped together, in the valley five hundred feet below the summit. Ewing had to break the terrible news to Mrs Hopkinson.

Hopkinson had been well aware of the urgent need of the Engineering Department at Cambridge for more accommodation, and a few days before his death had promised Ewing to open a subscription list for the north wing. His widow, aware of his intentions, wrote to the Vice-Chancellor on 13 October and jointly with her two eldest surviving children, Bertram and Ellen, offered to give £5,000 towards the construction of the new wing as a memorial to her husband and her son Jack, who, if he had lived, would have entered the department as an undergraduate in the Michaelmas Term. A syndicate prepared plans and estimates for the building, and the Public Orator read a letter of thanks to the three donors, declaring that

> No gift could be more noble in itself, more pathetic in its origin, none more appropriate in its purpose...The memory of the father and the son will be enshrined in the most fitting of Memorials, and will be permanently associated with the expansion of the scientific study of Engineering in Cambridge; while the names of the benefactors to whose generosity the Memorial is due will ever be retained in the grateful remembrance of the University.

Work on the new wing started as soon as the Easter Term was over and the building was completed in January 1900 at a cost of just over £5,500. It contained three laboratories, a lecture room for 120 and an attic floor with small rooms for research students. It was opened by Lord Kelvin and the Vice-Chancellor on 2 February. Cash donations amounted to about £1,800, and among many important pieces of equipment presented were a 5 ton testing machine, three dynamos, a 50 horse power steam engine and an 11 horse power gas engine.

The first buildings on the Downing site were started towards the end of 1901, and in the following summer the Department of Botany moved into its new quarters. Owing to the unforeseen expansion in the number of his students Ewing secured an increased allocation of the land and buildings vacated, and in addition to those promised him in 1897 he received the herbarium whose roof had been raised by Lyon, the private rooms of the Professor of Botany, ground on which a drawing office was built in the following year and the temporary use of a lecture-room.

Ewing was now a recognized authority on higher technical education. In 1901 he received honorary degrees from St Andrew's and Edinburgh Universities and was elected to membership of the Athenaeum. In the same year he sat on a Royal Commission which studied higher education in Ireland, and in 1903 represented the university at a conference convened by the National Association for the Promotion of Technical and Secondary education. Shortly before this, in November 1902, the parliamentary secretary to the Admiralty sought his advice on the training of engineer officers for the Royal Navy. The Board was engaged in a radical reorganization of the methods of entry and training of both officers and ratings, egged on by the explosive 'Jackie' Fisher, who was then Second Sea Lord. Fisher and Ewing soon became close friends, and early in 1903, at Fisher's instigation, the First Lord of the Admiralty, Lord Selkirk, offered Ewing the newly created post of Director of Naval Education, with a salary of £2,500 a year and an official residence in the Royal Naval College, Greenwich. Ewing felt it was his patriotic duty to accept the offer, and on 28 July he placed his resignation in the hands of the Vice-Chancellor, to take effect from 30 September.

Before leaving Cambridge Ewing obtained the appointment of a syndicate to consider the future of his department, as he had in mind certain reforms in the interests of his staff and his successor which he wished to see implemented before quitting the professorship. In particular he wanted to safeguard the position of his two demonstrators, whose tenure of office might, under existing regulations, have terminated with his resignation. The syndicate, of which he was a member, was appointed on 30 April 1903 and reported on 16 May.

The report described the growth of the department since Ewing's appointment in 1890. In twelve years the number of students had increased fivefold, and receipts from fees had risen from £546 to £5,005. The department was now managed by a triumvirate consisting of the professor and the two demonstrators, Lamb and Peace (who had succeeded Dalby), and there were eleven assistant demonstrators.[1] The syndicate contended that officers discharging functions as important as those of Ewing's two lieutenants were not properly described as demonstrators, nor were their status and responsibility adequately recognized, and they proposed the creation of two readerships for them, one in Electrical and one in Mechanical Engineering. The recommendations were endorsed by a majority of the General Board, but when they came up for discussion there was considerable opposition on the grounds that demonstrators in other departments were equally worthy of promotion and that all cases should be considered together. It was also suggested that lectureships might be more appropriate than readerships. Ewing argued that the Engineering Department now covered so wide

[1] W. Hartree (Trinity); C. E. Inglis (King's); W. S. La Trobe (St John's); T. Peel (Magdalene); A. Chapple (St John's); A. H. Peake (St John's); F. Grant (Queens'); J. W. Landon (Sidney Sussex); F. J. Dykes (Trinity); D. C. Highton (King's); H. Davidge (St John's).

a field that the real alternative to decentralization was the division of the work between independent professors, but if the Syndicate wished to keep the department under a single head, those in charge of the main branches of study should certainly have the status of readers. The general opinion of the Senate seemed, however, to favour the lower rank, and an amended report, published on 30 May, made this concession. To obviate any possibility of misunderstanding regarding the professor's responsibility for the whole department the Syndicate proposed to amend the Ordinances to read, 'The Department of Mechanism, including the Engineering Laboratory and the Mechanical Workshops, shall be managed by the Professor of Mechanism and Applied Mechanics; he shall have general control of the lectures and all other teaching in the Department; and he shall be responsible for the whole administration of the Department.' The report was approved without discussion on 11 June, and four days later Lamb and Peace were appointed to university lectureships in Electrical and Mechanical Engineering respectively.

Ewing set himself one more task before leaving Cambridge, and that was the provision of money for rebuilding the workshops, which he estimated would cost about £10,000. When his last report was published in June 1903 he was able to tell the university that £1,750 had already been subscribed and that an anonymous donor had promised to give one-quarter of the total sum required if the remainder were obtained from other sources.

Ewing left the Engineering Department in a very different condition from that in which he had found it. When Stuart resigned numbers were falling, the method of instruction was under severe criticism and the very existence of the department was in doubt. Now it was held in high esteem both inside and outside Cambridge, and Ewing's personal popularity was outstanding. The Sadlerian Professor expressed the general feeling when, at the discussion on the future of the department, he said it was difficult to think of it without Ewing, and it was only now that he was leaving them that members of the Senate were beginning to realize how much he had done for them. He received compliments and good wishes from all quarters, but the gesture which probably pleased him most was his election to an honorary life fellowship by his college.

Ewing was a very successful Director of Naval Education, and in the Coronation Honours of 1911 he was made a Knight Commander of the Bath. In the same year he married Ellen, the surviving daughter of his old friend and patron John Hopkinson, his first wife having died two years previously after a long illness. On the outbreak of war in 1914 he moved from Greenwich to the old Admiralty building in Whitehall and was placed in charge of the cypher-breaking team in Room 40. There, according to Winston Churchill, the First Lord, 'he made a contribution to the affairs of the Admiralty and to the fortunes of the State which might almost be called inestimable'.[1]

In 1916 he handed over responsibility for Room 40 to the Director of Naval Intelligence in order to become Principal of Edinburgh University. Three years later when, in the course of the post-war reorganization, a Board of Engineering Studies was established at Cambridge, Ewing together with A. B. W. Kennedy and two other distinguished engineers was elected an additional member. He resigned from his office of principal in 1928 and returned to Cambridge—'that delectable town', as he called it—where he bought a house in Adams Road overlooking the Rugby football ground. It was in these last years that he derived his greatest satisfaction from the honorary fellowship at King's. Further honours were showered on him, and in 1932, at the age of seventy-seven, he was President of the British Association at its York meeting. In the following year he was made a Freeman of his native city of Dundee, and on 7 January 1935 he died peacefully at his home in Cambridge.

[1] See A. W. Ewing, *The Man of Room 40* (London, 1939), ch. 9.

5 BERTRAM HOPKINSON

Professor of Mechanism and Applied Mechanics, 1903–1918

RAISING THE STANDARDS, 1903–1914

Ewing was succeeded as Professor of Mechanism and Applied Mechanics by Bertram Hopkinson, eldest son of the John Hopkinson who had died so tragically in August 1898.[1]

Bertram was born on 11 January 1874 and was three years old when the family moved from Manchester to London. He was educated as a day-boy at St Paul's School and soon made his mark there as a mathematician, winning a major scholarship to Trinity College, Cambridge, before he was seventeen. He did not go up to the university immediately, but continued his schooling and helped his father in the experiments connected with his research. He matriculated in 1893 and read for the Mathematical Tripos. He was sick for the examination in Part I and had to be content with an *aegrotat* degree, but when he took Part II in 1896 he was placed in the first division of the First Class.

After graduating he decided to enter the legal profession, where John Hopkinson's practice in cases involving patent rights ensured him useful introductions. He worked for a time in counsels' chambers and was called to the Bar in 1897. In the following year he joined the rest of his family in Switzerland for a short holiday before taking ship for Australia to conduct an inquiry on behalf of his firm, but the fatal accident on the Dent de Vesivi, which occurred after he had left them, changed the whole course of his life. He was at Aden when the news reached him, and he at once decided to abandon the law and carry on, as far as he was able, his father's unfinished work in the field of engineering and technological education. He went into partnership with his uncle Charles Hopkinson and his father's old assistant Ernest Talbot and worked on the design of electric tramways in Crewe, Newcastle and Leeds, but he continued to do some private research,

[1] See p. 124.

and in 1902 was awarded a Watt Gold Medal by the Institution of Civil Engineers for experiments on the electrolysis of pipes.

In 1903 Ewing resigned his professorship and Bertram, who knew him well, decided to apply for the post. He was only twenty-nine years old and had had no teaching experience, but he had already made a considerable professional reputation for himself, and Ewing, recognizing his outstanding ability, agreed to support his application. There were other candidates in the field, including J. B. Peace the Lecturer in Mechanical Engineering, but when the election took place on 14 November Hopkinson was chosen, and on the twenty-sixth of that month was formally admitted to the Professorship.

He had recently become engaged to Mariana Siemens, daughter of Alexander, third cousin and adopted son of Sir William Siemens, one of the three founder brothers of the great German firm of electrical engineers, and they were married on the last day of December 1903.

During the interregnum after Ewing's resignation the Engineering Department was administered by Peace, who afterwards served the new professor with the same loyalty as he had given to his predecessor. When Hopkinson assumed responsibility he made no sudden alterations in the curriculum. His decision to apply for the chair had been taken at short notice, and as he had embarked on his new duties without any clearly defined plans for the future he needed to see the existing system in action before deciding whether any changes were desirable. He had inherited his father's passion for research and it was his ambition to establish a school of research in engineering comparable to that in experimental physics in the Cavendish Laboratory. His interest in teaching developed rapidly in his new environment.

His first annual report to the Museums and Lecture Rooms Syndicate described the state of the department when he took it over. The new drawing office, lecture-room and Mechanical Laboratory had been completed during the previous summer, and for the first time for many years the space available for teaching was equal to the demands made on it. On the other hand, the equipment was in many respects inadequate for the number of students and much of it was out of date. Hopkinson repeated his predecessor's request for new workshops, and added a plea for an engine room at a further cost of £15,000. His staff consisted of

Lamb and Peace, lecturers, Inglis and Peake, demonstrators, and ten assistant demonstrators paid out of the fees.[1]

By this time the nineteenth-century system of personal teaching by private coaches was gradually being superseded by the appointment of college supervisors, but there were still not nearly enough of the latter available and a certain amount of coaching continued for about another fifty years. Ewing had allowed some of the coaches to give lectures in the department as well as private tuition in their homes but Hopkinson was opposed to this policy. However, he made an exception in the case of A. Chapple, principal mathematical coach to many generations of engineering students, whose position was so firmly established that he was allowed to continue his lectures in the role of an assistant demonstrator.

During the Long Vacation of 1904 Hopkinson made up his mind to raise the standard of the tripos course and demand from his students a greater knowledge of mathematics and its application to engineering problems. However, he did not rush matters, as he knew that his reforms would arouse opposition, and two years passed before he took any drastic action. He made one addition to the curriculum by arranging for C. T. Heycock of King's to give a course of lectures and laboratory work on the elements of inorganic chemistry and installed a small chemical laboratory in William Chapman's old house in Free School Lane. He himself lectured to the first-year men on applied mathematics and electricity.

In addition to his academic work Hopkinson continued his practice in the law courts, and in October 1904, to help with the preparation of his cases as well as with his private correspondence, he engaged as secretary Arthur H. Chapman, son of the Principal Assistant at the Fitzwilliam Museum and grandson of William Chapman the builder. Chapman, fresh from the Perse School at the age of sixteen, was a skilful draftsman and by adding typing and photography to his accomplishments soon made himself invaluable to Hopkinson and later to other members of the staff who were engaged on research. He was to serve successive heads of the department for half a century.

[1] J. F. Cameron, W. Hartree, W. S. La Trobe, T. Peel, H. Rottenburg, F. J. Dykes, J. W. Landon, F. Grant, A. Chapple and D. C. Highton.

Over the next few years Hopkinson built up a team of research students whose work was at first concerned mainly with the phenomena of gaseous explosions but was later extended to the more general effect of explosions and the impact of bullets on steel plates. He supervised his students very strictly, visiting them every day when he first arrived in the laboratory and following their work with the closest attention. Among them were many men who achieved great distinction in after life, including Harry R. Ricardo, who in his Presidential Address to the Institution of Mechanical Engineers nearly forty years later acknowledged his debt to Hopkinson and gave an interesting account of his methods.

Hopkinson was, I think, quite the most stimulating research leader I have ever met, with an almost uncanny perception combined with good judgement and a thoroughly practical outlook. Hopkinson's methods were by no means always orthodox; he believed in following step by step a logical and reasoned sequence, but only up to a point; if that looked like becoming too prolonged, then he would fall back on the principle of trying every bottle on the shelf, and if that did not achieve his end his next step was to try something really silly and see what happened. He taught me never to accept anything at second-hand unless it accorded with one's own common sense and experience; to be sceptical of one's own observations when they failed in this respect, and never to cling too long to a theory, however cherished.[1]

In the Lent Term 1906 Hopkinson joined the General Board of Studies as representative of the Special Board for Physics and Chemistry. On 16 March the Special Board for Mathematics, of which he was also a member, published a highly critical report on the Mathematical Tripos which expressed the view that in some respects the examination exercised an unsatisfactory influence on the course of study of the candidates. The order of merit and the great kudos attached to the position of Senior Wrangler encouraged the abler men to spend an excessive amount of time on detailed work in the less advanced parts of the subject, leaving too little for the higher branches. The less gifted mathematicians would, it was thought, receive a better education from a simpler and more general course, and this would be of particular value to those whose main object of study was the Mechanical or Natural Sciences Tripos, for whom the existing syllabus was unsuitable.

[1] *Inst. Mech. E. Proc.* CLII (1945), 143.

The examination itself was also considered unsatisfactory as the examiners were faced with the impossible task of setting questions whose primary object was to enable them to arrange the best candidates in order of merit while at the same time covering a syllabus intended to ensure a sound mathematical education for students of widely varying interests and abilities.

The solution proposed was a radical one. Part I of the tripos was to be an elementary examination which could be taken by the better students at the end of the first year and by the others at the end of the second. It would be a suitable preliminary both for the mathematicians who wished to proceed to the higher aspects of the study and for those students of engineering or physics who wanted a mathematical course before proceeding to the tripos in their chosen subject. Successful candidates would be arranged in three classes, each in alphabetical order, and the examination would not qualify for a degree. Part II, comprising Schedules A and B, was to be a much more advanced examination restricted to students who had already obtained honours in one of the triposes. Successful candidates were to be classed, again in alphabetical order, as Wranglers, Senior Optimes and Junior Optimes.

The report unleashed a torrent of argument and counter-argument, and opinion in the Senate was sharply and almost equally divided. The discussion took place on 3 May in a crowded Senate House, and there were many speakers. J. J. Thomson, the Cavendish Professor, supported the proposals. He said that the divorce between mathematics and physics in the tripos examinations was complete, and if a physicist wanted his mathematics tested by examination he was compelled to devote two years to hard study with little practical value attached to it before sitting for the tripos. The standard of the teaching of mathematics in schools was shamefully low, and a Part I designed for physicists and engineers might encourage the schools to raise their standards. The university should realize that the Mathematical Tripos was not the end of a student's career; the important part of an undergraduate's life was in front of him, and the university ought to aim at making the tripos course one which would best fit him for his future work. He compared his own post-graduate students with those who came to him from

American universities, very much to the advantage of the latter. There were no courses in America comparable in severity to the tripos, but although the Americans knew less when they arrived they generally retained a fund of energy and enthusiasm which the Cambridge men had exhausted by the time they sat for their examination. He supported the proposal to do away with the order of merit because he thought that all the triposes, and especially the Mathematical one, were too competitive and too difficult.

Hopkinson's contribution to the discussion showed clearly the lines on which he wanted his department to develop. The science of engineering, he said, must be the product of a union between mathematics and physics, and it was the mathematical engineer which Cambridge ought to produce. This was, in his opinion, the kind of engineer that employers looked for and often failed to find. Nine out of every ten of his students learnt all their mathematics within the walls of the Engineering Laboratory. Cambridge was the traditional home of mathematics, but most of his men never met the university's real teachers of the subject. Three or four came to him each year from the Mathematical Tripos and they were usually his best students. At present an engineering student had three courses open to him. He could either drop mathematics altogether and go straight to the Mechanical Sciences Tripos, or do one year's mathematics and follow it with two (or sometimes three) years of engineering, or do two or three years mathematics and sit for that tripos. The second course was the best professionally, but the men who took it had nothing to show for their first year's study. Most of his students dropped mathematics in order to get good places in their own tripos, and this largely accounted for the decline in the number of candidates for Mathematics. If the new scheme were adopted he would encourage his men to take Mathematics Part I before coming to Engineering, and he was confident that the new examination would soon be recognized by employers as a valuable qualification for an engineer.

An amended report, published on 29 May, summarized the proposals for reform in twelve recommendations, and these were all approved by small majorities on 25 October. The struggle, however, was not yet over. A third report, published on 13 November, contained the regulations for the new examinations and three recommendations for putting them into effect. It produced a flood of fly-sheets and letters to *The Times* and the discussion a fortnight later showed that the gulf between the reformers and their opponents was as wide as ever. Speeches on both sides were unusually bitter, and accusations of prejudice and bad faith were bandied about in a manner which proved that mathematics was still considered the lynch-pin of a Cambridge education, and that many of those who had devoted their lives to its teaching would fight to the last ditch rather than surrender to those who wished to change the traditional methods. When the resolutions were put to the Senate on 2 February 1907 non-residents came from all parts of the country to record their votes. All three Graces were non-placeted, but they were approved by 776 votes to 644, by 780 to 638 and 777 to 637. The battle was over at last.

While this great struggle was in progress proposals for the reform of teaching and examinations in the Department of Engineering attracted comparatively little attention and were approved without much open controversy. The report of the Special Board for Mathematics which embodied them was published on 12 March 1906, four days before that on the Mathematical Tripos. It defined the problem as follows:

> The number of Freshmen who annually commence to read for the Mechanical Sciences Tripos averages about 75. Of these a large proportion have a totally inadequate knowledge of Mathematics and Mechanics, evidenced in many cases by inability to pass the Previous Examination in the Additional Subjects on entering the University. Many cease to read for the Tripos at the end of their first year, but there still remain a proportion who are unsuited to take an Honours course which necessarily requires a considerable mathematical equipment. The result is seen in the Tripos Lists, from which it appears that out of 41 candidates in 1904, 19 failed to obtain Honours. In 1905 there were 45 candidates, of whom 18 failed to obtain Honours.

The Board believed that most of the men who failed the tripos would have been better served by a course involving less advanced mathematics. They were determined not to lower the standard of the Mechanical Sciences Tripos, but decided instead to widen the scope of the Special Examination in Mechanism and make it acceptable to the professional institutions as a qualification for associate membership. Candidates for the tripos would be required to take a Qualifying Examination in Mathematics and Mechanics before the end of their fourth term and those who failed to pass would be diverted to the course for the Ordinary Degree. Men who sat for the tripos but failed to obtain Honours would only be admitted to the Ordinary Degree if, in the opinion of the examiners, they would have obtained a first class in the 'Special'. In the tripos itself Part II, for which there had rarely been more than one candidate each year, was to be abolished. The examination was in future to consist of eleven papers in two groups, A and B, one containing questions within the capacity of any candidate deserving honours and the other containing questions of greater difficulty or wider range. In place of the practical examinations, which had become very difficult to conduct owing to the large numbers of candidates, examiners were to take into consideration laboratory and drawing office work as evidenced by duly attested note books. These changes were embodied in Regulations and Schedules.

The new regulations for the tripos were approved on 24 May 1906 and those for the 'Special' on 14 March 1907. The latter provided that a candidate should not be placed in the first class unless he satisfied the examiners in the three compulsory subjects and in two of the three alternative ones, viz. Heat and Heat Engines, Strength of Materials and Elementary Theory of Structures and Electricity.

These reforms produced the results which Hopkinson aimed at, and fears that the higher standards would frighten away a large proportion of the students proved unfounded. For the next few years numbers dropped slightly, but better men were attracted to take the place of those excluded by the Qualifying Examination. The Honours course gained steadily in reputation, and though the course for the Ordinary Degree was never entirely satisfactory it was a great improvement on anything which had been done for the engineering poll-men in the past. The next seven years marked a period of steady progress which lasted until the outbreak of the First World War brought the normal work of the university to a sudden halt.

Hopkinson devoted a great deal of personal attention to the teaching of his undergraduates, by whom he was held in considerable awe. Two or three times every term he inspected their laboratory notebooks and discussed the experiments with each student individually before initialling them. He had an unerring eye for slipshod work or for any attempt to gloss over something which had been imperfectly understood, and his interest had a marked effect on the teaching. The standards for the Honours Degree rose steadily and a new spirit imbued the poll-men, whose work also showed a corresponding improvement.

He also found it necessary to intervene personally in the running of the workshops, where discipline among the workmen had become extremely lax. They came in at all hours and no one kept a proper check on their attendance. Ewing had realized this, and when Hopkinson's appointment was announced he warned them that they would have to mend their ways. Many of the teaching staff made use of the workshop facilities, but no one was personally responsible for them or for the workshop instruction of the undergraduates, who were sent by their tutors to make arrangements direct with the instructor craftsmen. Test jobs had to be produced for the practical examination, and it was well known that some of the workmen were willing, for a consideration, to 'oblige the young gentlemen' by finishing them off.

Hopkinson found this situation intolerable and for a time used to appear at the workshops at 8 a.m., when they were supposed to open, with his watch in his hand. He soon began to look around for someone to take charge of the workshop activities and about 1909 G. F. C. Gordon, M.A., of Trinity (known to generations of undergraduates as 'Foundry Freddie') assumed this duty, though he held no official position until the office of Superintendent of the Workshops was revived by Inglis in 1920. His appointment was bitterly resented by some of the craftsmen, who insisted on maintaining their right of appeal to the professor when disagreements arose.

After the institution of the Qualifying Examination numbers in the department stabilized for a time at about 260. In comparison with former years rather fewer men took the Tripos and there was a corresponding increase in the number of candidates for the Special. In March 1908 Peace resigned his lectureship to become University Printer, though he continued for some years to lecture and supervise, and Inglis succeeded him as Lecturer in Mechanical Engineering. A third demonstratorship was created and J. W. Landon and T. Peel joined A. H. Peake in this grade.

In March 1909, after long and sometimes acrimonious discussions, the regulations for the Ordinary Degree examination were again recast. The General Examination remained much as it had always been, but the number of Specials was raised to eighteen and they were divided into nine groups. Group VII consisted of Mathematics and Mechanism and Applied Science and Group V contained all the other scientific papers—Chemistry, Physics, Geology, Botany, Physiology, Zoology and Agricultural Science. Candidates could either take the General Examination and one Special, as previously, or two Specials from different groups. In the discussion on the report Hopkinson said he knew little about the working of the General Examination but he did claim some knowledge and experience of the poll-men. Although Engineering was the most difficult of the Specials the number taking it increased continuously, and this was probably because in the Engineering Department the poll-men were catered for as much as the honours men. They had separate classes and the teaching was organized in the same way and with the same care for both types of students. Among those who had taken the Ordinary Degree were many who had done extraordinarily well in after life as engineers in all parts of the world. Hopkinson thought that other departments ought to pay more attention to the education of the men who were not candidates for honours, and believed that the Specials could be improved with that end in view.

The reform of the Mathematical Tripos produced the expected repercussions on the engineering students, and in the Michaelmas Term 1911 a class in engineering was formed for the benefit of twenty freshmen who decided to take the new Part I examination in mathematics before reading for the mechanical sciences.

By this time the number of undergraduates in the department was again increasing and the need for more accommodation was making itself felt. In 1912 Hopkinson obtained permission to use part of the accumulated savings of the department to pay for the erection of a new building in place of the old Perse headmaster's house. This was accordingly demolished and in its place a block was built along the Free School Lane frontage joining the Engineering and Cavendish Laboratories. It had three floors and a basement, each with an area of about 1,800 square feet. The Electrical Laboratory was transferred to it, and as a result substantial improvements were made to the engine room and boiler house in the old buildings. The total cost, including a new heating system for all the departmental buildings except the workshops and drawing office, was in the region of £4,000.

The year 1912 also saw the revision of the regulations for Advanced Students. Of 101 admitted to the courses of advanced study between 1896 and 1909 only 35 had reached the required standard in the tripos, while 18 failed and 48 did not present themselves for examination. Of the six students admitted for advanced study in the mechanical sciences two passed, one was rejected and three were not examined. Under the new regulations men admitted for research were to be known as Research Students, and those admitted from other universities to read for tripos examinations were to be called Affiliated Students. The Statutes were amended in February 1913 to give effect to these changes.

The regulations for the Special Examination in Mechanism and Applied Science, which had not been changed substantially since 1903 were also revised in 1913. The papers were re-arranged in two parts. Part I was designed to cover the minimum knowledge of mathematics, mechanics and materials needed by an engineer. Part II was more specialized and more advanced.[1] The examination was renamed 'The Special Examination in Engineering Science'.

THE UNIVERSITY AT WAR, 1914–1918

The outbreak of war in 1914 did not find the university entirely unacquainted with military problems. As early as October 1900 one hundred and sixty-one members of the Senate had signed a memorial declaring that the time was ripe for the organization of instruction in Military Sciences in the university, and within a week a syndicate, of which Ewing was a member, had been appointed to look into the matter. The syndicate was not able to complete its investigations until January 1904 as it had to wait for the findings of a Committee on Military Education set up in 1901 by the Secretary of State for War to consider the problem on a national basis. In a Blue Book issued in May 1902 this committee reported that the military authorities were 'practically

[1] The Part II papers were on Mechanical Drawing, Mechanics of Machines, Heat Engines, Theory of Structures, Applied Electricity, Hydraulics and Surveying, the last subject being optional. Candidates were also required to satisfy the examiners of their proficiency in workshop practice.

unanimous' in regarding the universities as potential sources of supply for desirable army candidates. A permanent Advisory Board for Military Education was established by Parliament and in due course regulations were issued for granting commissions to university graduates. These regulations stipulated that candidates must have graduated at a residential university approved by the Secretary of State, be between twenty and twenty-five years of age, have obtained a certificate of military proficiency and passed an examination in military subjects. Commissions were to be assigned every six months to groups of approved universities by a Nomination Board on which the War Office would be represented. Cambridge candidates who had obtained first class honours in a tripos would be granted one year's seniority if selected.

In March 1904 a Board of Military Studies was established by the Cambridge Senate to arrange for the instruction of candidates for commissions, to nominate for commissions those who had qualified, to organize teaching in military subjects and to establish a liaison with the War Office and with other universities on matters connected with military studies. C. T. Heycock of King's was one of the original members and Hopkinson joined the Board in 1909. In that year Military Subjects was accepted, though not without opposition, as another Special Examination for the Ordinary Degree.

As the threat of war with Germany grew stronger shortages in the Territorial Forces caused increasing concern, and in May 1914 T. F. C. Huddleston and C. T. Heycock published an article in the *Nineteenth Century and After* proposing that university students should be required to reach a certain standard in military training in order to qualify for the first degree. Oxford and Cambridge in particular were considered to have the responsibility for educating their members in the duties of citizenship, and it was suggested that they could set an example by diverting the energies of the students from excessive athleticism into more patriotic channels after the manner of Switzerland, where marksmanship was as popular with students as football. Proposals were also made for establishing in the public schools and other centres cadet corps authorized to award proficiency certificates which would be required from all candidates for the civil service, the Bar, the medical and other professions. A memorial embodying the views expressed in the article was signed by nearly two thousand people from all parts of the country, and a copy bearing the names of 186 members of the Cambridge Senate was sent to the Vice-Chancellor on 23 May 1914. A counter-memorial, carrying 163 names, expressed the contrary view 'That it is highly undesirable that a University should by its own authority require any form of military service from any of its members'. Among the signatories of the second document were those of Hopkinson, Lamb and J. J. Thomson.

Hopkinson's opposition to the proposal that military qualifications should be required for admission to a Cambridge degree in no way implied that he objected to voluntary military training in the university. On the contrary, he did everything he could to encourage it. His father had been a pioneer in founding the London Electrical Engineers in the old Volunteer Force, and Bertram took command of the Cambridge detachment of this body soon after taking up his professorship. When the Officers Training Corps was established in 1908 most of the undergraduate 'sappers' transferred to the new organization. Hopkinson was put in command of the Fortress Company R.E. in the university contingent with the rank of major and Landon and Inglis became his second lieutenants.

Hopkinson also played an extremely important part in the national preparations for war by carrying out research on various matters of interest to the service departments. His investigations lay mainly in three fields—first, the study of metals, particularly with regard to the magnetic properties of alloys, the elastic properties of steel and other metals and the phenomena of elastic hysteresis; secondly, the development of gas and other internal combustion engines; and, thirdly, the science of flame and explosion. In 1908 he designed a torsionmeter for the Admiralty which was installed by John Brown Limited in the battle cruiser *Inflexible*. In 1911 he began a series of demolition experiments on the C.U.O.T.C. range in Grange Road, and in 1913 he presented a paper on 'Bullets' to the Royal Society. Later in the same year he began a series of researches on detonation and the effect of explosions in contact with steel plates. He also gave expert evidence before a War Office Committee on Wireless Telegraphy.

Britain declared war on Germany on 4 August 1914, and two days later the Vice-Chancellor, in his capacity as Chairman of the Board of Military Studies, appointed a special War Committee to deal with the selection and recommendation of applicants for commissions. It sat at first at the High Table in Corpus Christi College, while voluntary workers occupied the body of the hall and Boy Scouts from the 1st (Sea Scouts) and 5th (Perse School) Troops acted as orderlies. When the new academic year opened in October the Cambridge O.T.C. was divided into two units, infantry and medical, and a regular system of training on three afternoons a week was organized. A military hospital was set up in Neville's Court in Trinity College, and later transferred to the King's and Clare playing fields where the University Library now stands. By the end of the first year of the war the university had shrunk to a third of its former numbers and by the Easter Term of 1916 there were only 575 undergraduates in residence as against 3,263 at Michaelmas 1913. In the Engineering Department only a handful of students remained—26 in the Michaelmas Term of 1915 and 17 a year later. All the teaching there was carried out by Lamb and Peel, while Peace supervised the workshops. By 1917 these were fully employed on government contracts, and the number of men and boys working in them had risen from the peacetime establishment of 12 to 33. To meet an urgent national need F. J. Dykes undertook the manufacture of gauges and was so successful that he was subsequently made Technical Manager of the Newall National Gauge Factory at Walthamstow. To cope with the new demands a considerable amount of new equipment was installed, but the work done was paid for by the government and the receipts were sufficient to keep the department in a sound financial condition during the war and to provide a substantial reserve for future developments.

Throughout the war years those members of the university staff who remained in residence gave much thought to the problems which would have to be tackled when peace was restored. A memorandum published in 1916, of which Dykes, Lamb and Peel were signatories, recommended an increase in the length of term at the expense of the Christmas and Easter vacations. the simplifying of the degree examinations, and the institution of short courses of study leading to a certificate. A syndicate was appointed to make a comprehensive study of the whole university examination system, and its report, published in May 1918, led eventually to important changes in the engineering courses.

Hopkinson played no part in this planning for the future of the university. He was mobilized as soon as the war started, and dropping all his other interests concentrated his efforts entirely on military matters. Leaving Lamb as his deputy in charge of the department he went as an instructor to the School of Military Engineering at Chatham, retaining his Territorial Army rank of major. Later he joined Ewing's cypher-breaking team in Room 40 at the Admiralty, but though he displayed a remarkable flair for this highly specialized activity it never occupied more than a portion of his time. The work on explosions which he had been doing at Cambridge directed his thoughts towards the protection of warships against mines and torpedoes and he played an important part in the development of 'blisters' on the hull with a structure inside them to absorb part of the energy of explosion. In July 1915 he was appointed to a panel of Admiral Fisher's Board of Invention and Research. His pre-war work on bullets led him to the design of bombs for use by aircraft and thence to the equipment of aircraft generally. In November 1915 he accepted a post in the Department of Military Aeronautics and organized a programme of research on bomb and gyro sights, guns and ammunition which was carried out at Cambridge, Farnborough and elsewhere. In the spring of 1916 he opened an experimental station at Orfordness which from that time became one of the main centres of his activities. Much of the work on explosives at Cambridge was taken over for him by H. W. Phear,[1] great-nephew of the Master of Emmanuel who had befriended Stuart.

Hopkinson's work was becoming even more closely related to the technicalities of flying, and although about twice the age of the average pilot he decided to learn to fly. He obtained his 'wings' and though too old to make a really good pilot he knew his limitations and took no unnecessary risks. He flew constantly between London, Orfordness, Farnborough and France, and although some of his friends felt that his life was too valuable to be risked in this fashion there is no doubt that his experience as a pilot aided his judgement in aeronautical matters

[1] Mr Phear was on light duty after a wound received in France. After the war he joined the teaching staff of the department, and lectured there until his retirement in 1960.

and immensely increased his influence and prestige with the officers at the experimental stations. He did much valuable work on the development of flying by night and in bad weather and on navigation in clouds. The testing of aeroplanes was put under his direction and units under his orders were established on Martlesham Heath and at Grain. In June 1918 he was appointed Deputy Controller of the Technical Department and later in the year was promoted to colonel. For his work on the underwater protection of ships he was made a Commander of the Order of St Michael and St George. On 26 August 1918, while flying from Martlesham Heath to London in a Bristol Fighter he ran into bad weather and while trying to find his way through thick cloud crashed and was killed. His body was brought to Cambridge and buried with military honours in St Giles cemetery after a service in King's College chapel which was attended by many distinguished mourners.

Ewing edited his scientific papers and prepared them with a biographical memoir which sketched vividly his appearance, personality and achievements.

Tall, of a commanding presence, with immense physical strength and energy, with ripe engineering experience and great originality of mind, of a perfect and unruffled kindliness and severity, he commanded respect and confidence in all those who worked with him. His character was in keeping with the external attributes of manliness, without a trace of self-seeking, without a thought that was not directed to the country's good, obviously ready to make any personal sacrifice, and as obviously of a strong, independent and fearless personality, he was born to command; and command he did...

The experiments of the Cambridge laboratory were of high interest in themselves and in their bearing on engineering practice. But to Hopkinson they were more; one may say they constituted an apprenticeship for the culminating work of his life, the work of the last four years. The war gave him an opportunity such as he did not have before. Into it he threw all his inventiveness, all his initiative, his untiring energy, his power of organization, his unrivalled capacity for getting the best out of himself and out of others. No worker rejoiced more in his work or accepted its call with more absolute renunciation. He was amazingly aloof from any consideration of private advantage or personal convenience. The strain was immense: the pressure of claims on his attention was continuous, but it never seemed to ruffle his serenity nor impair the soundness of his judgement. Many will mourn him as a trusted friend, but only those who knew something of what he did in the war can have a right idea of the magnitude of the nation's loss.

6 CHARLES EDWARD INGLIS

Professor of Mechanism and Applied Mechanics, 1919–1934
Professor of Mechanical Sciences 1934–1940
Head of the Department of Engineering 1919–1943

BACKGROUND TO THE PROFESSORSHIP, 1875–1919

Charles Edward Inglis came of an ancient Scottish family whose estate of Auchindinny lies on the North Esk about nine miles south of Edinburgh. It had been bought in 1702 by John Inglis, Writer to the Signet, and was occupied by successive members of the family until Charles's father, Alexander Inglis, came south about 1870 to practise as a doctor in Worcester. He let the estate to tenants during the remainder of his lifetime, but on his death it was reoccupied by his eldest son, John Alexander, who became King's Remembrancer for Scotland.

Charles Edward, Alexander's second child, was born in Worcester on 31 July 1875. His mother died eleven days later, and after a short time Alexander moved to a larger practice in Cheltenham, where his sister Jane kept house for him and looked after the two children. He married again in 1883, but suffered a second bereavement a year later, and Jane resumed her duties as housekeeper. When the boys were old enough they went to Cheltenham College, where Charles found many interests besides his school work, particularly in astronomy and model making. His father encouraged these hobbies and allowed him to mount a telescope in the loft over his stables and fit up a small room in the house as a workshop. In 1891 Alexander married for the third time, and his new bride brought into the home a greater sense of warmth and happiness than it had ever known before. There were two children by this marriage and the elder boys became devoted to them and their step-mother.

In 1894 Charles won a minor scholarship to King's College, Cambridge. He read Mathematics and graduated in 1897 as twenty-third wrangler, after which he spent a fourth year reading Engineering under Ewing, obtaining first class honours in the Mechanical Sciences Tripos of 1898.

He had been a puny infant whose life had at first been despaired of, but he was now a young man of exceptional stamina. He loved long walks and long-distance running and revelled in what he sometimes described as 'the luxury of physical exhaustion'. He would probably have been awarded his Blue for running if he had not pulled a muscle in the crucial race. Like his two predecessors in the chair of Mechanism, Hopkinson and Ewing, he later became an enthusiastic mountaineer.

After going down from Cambridge he went as a pupil to the London office of the distinguished civil engineer Sir John Wolfe Barry. Barry was a born teacher and his pupils received a first-class training. They were taken on for a five-year course at a fee of five hundred guineas, spending their first year in London and the rest of the time on building sites. Inglis did his preliminary training in the drawing office under the chief designer and distinguished himself by particularly good work on the drawing board. He was then appointed to the staff of the Resident Engineer, Alexander (later Sir Alexander) Gibb, who was working on an extension of the Metropolitan Railway between Whitechapel and Bow. For the greater part of its length the cutting ran under the Whitechapel, Mile End and Bow roads, but traffic, including the trams, was kept moving over the whole route throughout the period of construction. Nine overline bridges had to be built, as well as a seven-span viaduct and four underline bridges, mostly on the skew over railways. Inglis was employed on the design and construction of these bridges, an experience which was of great value to him in later life. However, he did not stay with Wolfe Barry long enough to see the opening of the railway, as after completing two years of his apprenticeship he decided to make his career in the academic world.

In October 1900 he entered for one of the scholarships offered by King's College to its graduates to help towards their professional training, which entailed the submission of a dissertation. The subject he chose was not derived from his practical experience in civil engineering; he had become interested in the theory of mechanical vibration, on which he later became a leading authority, and he submitted a paper on 'The Balancing of Engines'. The referees were Ewing and W. H. Macaulay, both Fellows of the college, and scholarships were awarded to Inglis and one of the other candidates. It is believed that this was the first fellowship awarded by any Cambridge college in recognition of the Mechanical Sciences.

Inglis was already working in the university as Ewing had engaged him as an assistant demonstrator at the beginning of the previous Lent Term. He was also giving supervision for his college and on winning the scholarship was appointed an Assistant College Lecturer. This entitled him to the use of a room, but he did not occupy it for long as before the end of the year he married Eleanor Mary Moffat, daughter of a colonel in the South Wales Borderers, whom he had met while on holiday in Switzerland. In the Michaelmas Term he took over from Peel the lectures on Mechanics for second-year tripos candidates and in July 1903 succeeded him in the established post of demonstrator.

Inglis was never happier than when he was lecturing, and Hopkinson, who became head of the department in October, soon recognized his outstanding ability as a teacher. In the years which followed he gradually built up the number of his courses until he was carrying a greater teaching load than any other member of the staff. His subjects included statics and dynamics, the theory of structures, materials and drawing. In the Long Vacation he gave courses on the balancing of engines, girder design and reinforced concrete. In 1906 he was one of the Examiners for the tripos and thus became an *ex-officio* member of the Special Board of Mathematics at the time when the two revolutionary reports on the triposes were being drafted.[1]

He did not, however, devote the whole of his time to teaching as Hopkinson encouraged him to pursue his research interests. He continued the work on vibrations which had formed the basis of his fellowship dissertation, and in 1910 co-operated with Hopkinson in a tract entitled 'Vibrations of systems having one degree of freedom'. This was the first of a series of papers on engineering problems published by the Cambridge University Press. In the following year Inglis published a paper on the balancing of the four-cylinder marine engine, and in

[1] See pp. 132–136.

1913 another on stresses in a plate due to the presence of cracks and sharp corners which was perhaps his most important contribution to engineering science.

When Hopkinson took over the Engineers Fortress Company in the Officers Training Corps Inglis became his second-in-command, and it was to occupy the Engineering Section of the Company on field days that he designed a portable tubular bridge for crossing the mill stream on Coe Fen. After a time several teams were formed within the section, and a bridging competition established for which Hopkinson presented a challenge cup.

On the outbreak of war Inglis was commissioned in the Royal Engineers and appointed to the War Office, where from 1916 to 1918 he was in charge of the department responsible for the design and supply of military bridges. His own bridge was adopted by the army and proved its worth when tanks went into action on the Western Front in 1917. He was awarded the O.B.E. for his services and promoted to the rank of brevet major. When the war ended he resumed his post in the Engineering Department, and on 25 March 1919 was elected to the professorship made vacant by Hopkinson's untimely death. He was then forty-three years old.

POST-WAR RECONSTRUCTION AND THE MOVE TO SCROOPE HOUSE, 1919–1922

On his return to Cambridge after the war Inglis found the university facing one of the gravest crises in its history. For four years the normal entry of undergraduates had been almost entirely suspended; no permanent buildings had been erected, no new equipment bought, and expenditure on maintenance had been kept to a minimum; many of the teaching staff had died during the war and few young men had been trained to replace them or those who had retired. Even to resume the work of teaching and research where it had been broken off in 1914, and with the same number of students, would have presented a herculean task; but far more than this was required. A high proportion of the young men who would have come up between 1914 and 1918 had been killed, and many of the survivors were obliged to forfeit the advantages of a university education in order to earn their living, but

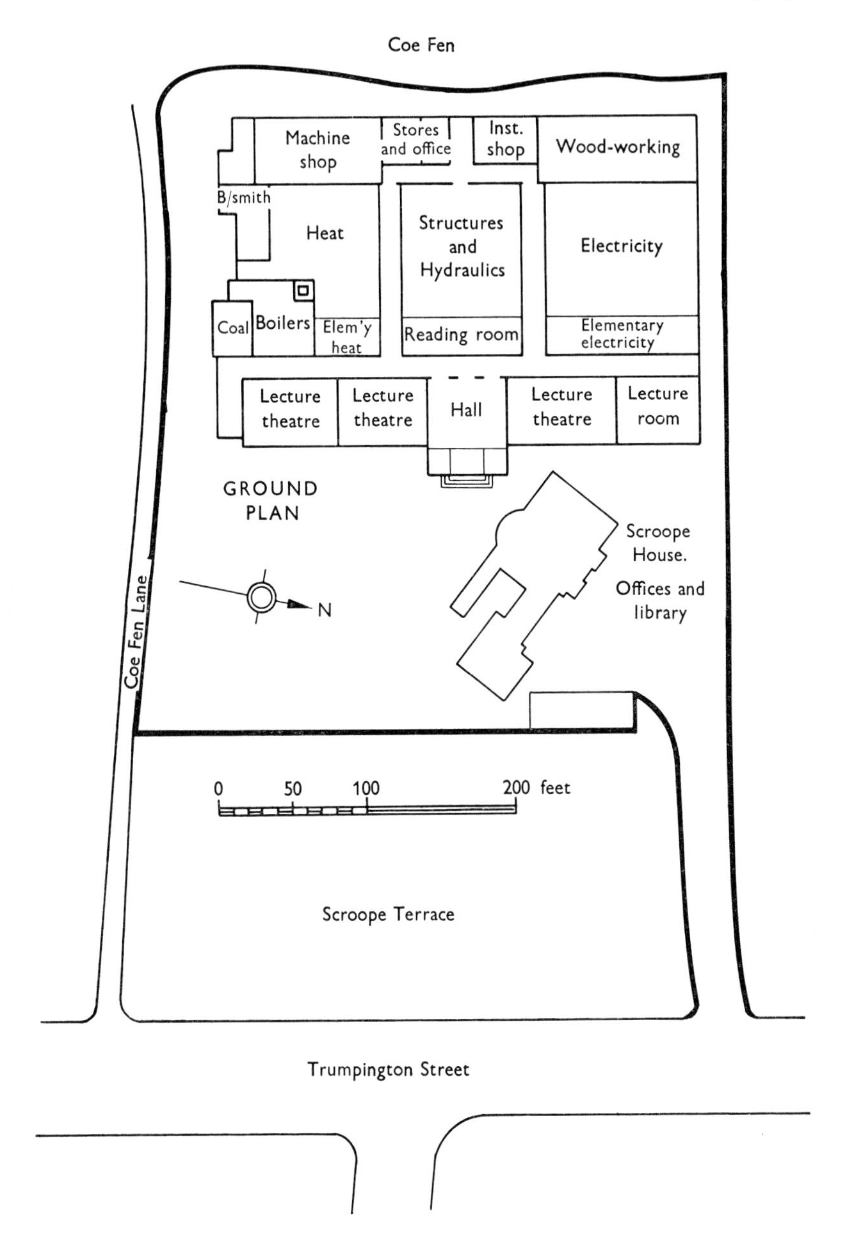

The original plan for the Inglis Building, 1919.
Architect: F. W. Troup, F.R.I.B.A.

enough remained to make the matriculations in 1919 and 1920 the highest on record.

The crisis was tackled with determination and resource. Emergency regulations were passed enabling men released from military service to graduate after two years instead of three, and the colleges provided extra accommodation by increasing the numbers living in and by allowing more men to go into lodgings in the town. The chief difficulties arose in the scientific departments, where the students required laboratory accommodation and apparatus on a scale which had never been foreseen. In some departments the fees did not even cover the direct expenditure incurred by the students, so the rise in numbers added to the annual deficit instead of increasing the net revenue. More modern laboratories and much new equipment were essential, but neither the university nor the colleges could provide the money to pay for them.

The plight of the graduate staff was even more serious than the material shortages. Money had fallen in value and pre-war stipends no longer provided a living wage, nor was there any pension scheme to mitigate the hardships of retirement. A new deal for university teachers was essential if existing staffs were to be retained and new members recruited, but for this also money was lacking.

Cambridge was not the only university to find itself in difficulties. Oxford too was on the verge of bankruptcy, and it was obvious that the provincial universities, which had long been in receipt of grants from the Treasury, would require much larger allocations than they had been receiving before the war. In the autumn of 1918 the government decided to make a comprehensive survey of the needs of higher education throughout the country. The President of the Board of Education, H. A. L. Fisher, obtained statements of the post war requirements of their universities from the Vice-Chancellors, and when these had been studied the Cabinet agreed on an important change in policy. In future every University receiving state aid was to get it in the form of a single inclusive grant instead of in separate grants to individual departments, and a new standing committee, to be known as the University Grants Committee, was created to advise on the sums to be allocated.

The minister's request for a statement of needs was received in Cambridge on 6 November 1918, five days before the signature of the armistice, and was answered by T. C. Fitzpatrick of Queens' who, in the absence of A. H. Shipley in America, was acting as Vice-Chancellor. He quoted from a forecast of post-war needs which had been drawn up by the General Board in 1916, and expressed his personal opinion that Cambridge should receive a substantial grant from the Exchequer 'provided that the conditions under which it is given do not interfere with the autonomy to which throughout the long history of the University the Senate has always attached the utmost importance'.

When Shipley returned from America in March 1919 the need for financial assistance was so urgent that he sent a deputation to London to put the case for an emergency grant to the minister in person. Fisher's reaction was sympathetic, but in a letter dated 16 April he pointed out to the Vice-Chancellor that Cambridge was in quite a different category from the universities which had received state aid before the war. Their grants had been agreed upon after their revenues and expenditure had been fully investigated, but no such inquiry had been made in recent years into the resources of Cambridge. The government would not feel justified in making Cambridge a grant from parliamentary funds, 'except on the condition that in due course a comprehensive inquiry into the whole resources of the University and its Colleges, and into the use which is being made of them, shall be instituted by the Government'. If this condition were accepted by the Senate the University Grants Committee would be instructed to submit recommendations for an emergency grant during the current financial year 'to meet the immediately urgent needs of salaries and maintenance'.

Fisher knew well that the prospect of a Royal Commission would not be welcome, and to allay misgivings he committed himself to a statement of his own attitude towards the principles involved.

> No one appreciates more fully than myself the vital importance of preserving the liberty and autonomy of the Universities within the general lines laid down under their constitution. The State is, in my opinion, not competent to direct the work of education and disinterested research which is carried on by the Universities, and the responsibility for its conduct must rest solely with their Governing Bodies and Teachers. This is a principle which has always been observed in the distribution of the funds which Parliament has voted for subsidizing University work; and so long as I have any hand in shaping the national system of education I intend to observe this principle.

This correspondence was put before the university in a report by the Council of the Senate dated 5 May 1919 and discussed eight days later. The report was approved without opposition on 31 May and the Vice-Chancellor was authorized to tell the minister that the university would welcome a comprehensive inquiry into its financial resources and would give any assistance in its power. Pending such an inquiry the government's attention was to be drawn to the urgent need for an emergency grant. A Joint Finance Committee was formed by the Council, the Financial Board and the General Board of Studies to conduct the negotiations with the government. The chairman of the University Grants Committee visited Cambridge in August and on his advice an emergency grant of £30,000 was approved for the year 1919–20.

The Royal Commission 'to enquire into the financial resources of the Universities of Oxford and Cambridge and the Colleges and Halls therein, into the administration and application of these resources, into the government of the Universities and into the relations of the Colleges and Halls to the Universities and to each other and to make recommendations' was appointed in October 1919. Its chairman was H. H. Asquith, and among the Cambridge representatives were Sir Horace Darwin, G. M. Trevelyan and Miss Clough of Newnham.[1] In the course of its investigations sixty-six meetings were held and over ninety witnesses examined.

In April 1922 Fisher announced in the House of Commons that the government had agreed that Oxford and Cambridge would in future be added to the list of universities in receipt of annual grants-in-aid from the Treasury, and that subject to parliamentary approval it was proposed that in the current year the grant should remain at £30,000 plus the £8,500 already approved for the Medical Department.

The report of the Royal Commission was published at the same time. It contained a detailed statement of the income of the two universities and their colleges,[2] and proposed for Cambridge a system of graduated taxation of the colleges in aid of the university similar to that already in force at Oxford. Stipends were recognized as being quite inadequate

[1] The members of the Commission were H. H. Asquith, Baron Ernle, Baron Chalmers, G. W. Balfour, Sir John Simon, Arthur Henderson, E. G. Strutt, T. B. Strong, Sir Howard Frank, Sir W. M. Fletcher, Sir H. A. Miers, Sir Horace Darwin, Sir J. H. Oakley, Prof. W. L. Bragg, G. M. Trevelyan, Miss E. Penrose, Prof. W. G. F. Adams, H. K. Anderson, Miss B. A. Clough, H. M. Cobb, M. R. James, A. Mansbridge, Prof. A Schuster.

[2] The totals were £824,710 at Oxford and £719,554 at Cambridge.

and a pension scheme as an urgent requirement. Some of the teaching staff were stated to be seriously overworked, with no leisure for study and research, and a considerable increase in their number was needed. There was also a requirement for studentships for graduates to enable them to pursue advanced studies in the university, and reserved fellowships in the colleges were advocated for university lecturers and demonstrators. Considerable sums were suggested as grants for the non-collegiate bodies, women's education and extra-mural work. For Cambridge an emergency grant of £302,500 was proposed for buildings and sites to make up for eight years' gap in building and maintenance during and since the war.

On the whole this report was well received by the Senate, and in May 1922 the Council published a resolution welcoming the immediate appointment of a Statutory Commission to put the proposals into effect.

While the university was battling with these major problems of finance and administration Inglis was putting the Engineering Department back on to a peacetime footing. On 19 May 1919, less than two months after his election to the professorship, the Syndicate on Engineering Studies, of which he, Peace and Lamb were members, issued its third report. It recommended the termination of dual control over the departmental teaching by the Special Boards for Mathematics and Physics and Chemistry and the establishment of a separate Board of Engineering Studies. This was approved on 17 June and the new Board held its first meeting on 2 December. The Vice-Chancellor was *ex-officio* chairman and J. F. Landon, the lecturer in Mechanical Engineering, was appointed secretary.[1] Four external members were elected—Sir J. A. Ewing, Sir A. B. W. Kennedy, Sir J. P. Griffiths and F. D. Docker. The Vice-Chancellor presided at the Board's meetings until March 1921, when he was relieved of this and other duties by Grace of the Senate, and in the following October J. F. Cameron was elected chairman—a post which he occupied for the next eighteen years.

The first act of the Board of Engineering was to set up a committee

[1] The original members were the Vice-Chancellor, the Professors of Mechanism and Aeronautical Engineering, the Lecturers in Mechanical and Electrical Engineering, Professors Sir W. J. Pope (Chemistry), Sir J. Larmor (Mathematics), E. S. Prior (French), Sir E. Rutherford (Experimental Physics), Sir R. T. Glazebrook, Dr W. L. Mollison, G. I. Taylor and J. F. Cameron.

to consider and report on possible changes in the Mechanical Sciences Tripos and on the introduction of post-graduate specialized courses to replace the existing system of 'B' papers. Other committees reviewed the regulations for the Qualifying Examination and considered the possibility of improving the course on surveying, in which Inglis took a strong personal interest.

No major alterations were in fact made in the tripos until 1924, but in 1920 the regulations for the Ordinary Degree were completely recast. The General and Special Examinations were abolished and Examinations in Principal and Subsidiary Subjects substituted. Two separate examinations in Engineering, known as Engineering I and II, ranked as principal subjects.[1]

In the Michaelmas Term 1918 there were 37 students in the department. When Inglis assumed the professorship in the Easter Term 1919 there were 283 full-time students and 95 naval officers attending short courses of lectures and laboratory work. In the following October 553 students were registered for the tripos and 109 naval officers were taking special courses. A year later the students totalled 808—a figure which was not reached again until 1959. To provide temporary accommodation for them three wooden army huts were erected on vacant courts near the old buildings. Two were used as lecture-rooms and one as a laboratory, and in addition the Small Examination School was lent to the department as a temporary drawing office.

During the last years of the war practically all the teaching had been given by Lamb and Peel, but by October 1919 Inglis had succeeded in building up a teaching staff of 38, 11 of whom had lectured or demonstrated in the department before the war.[2] In June 1919 a University

[1] The subjects were divided into two groups. Group I consisted of 22 principal and 13 subsidiary subjects, mainly in the arts, and Group II of 14 principal and 11 subsidiary subjects, mainly scientific. Candidates were required to qualify in five subjects, at least three of which had to be principal, and one of the five, either principal or subsidiary, had to be taken from each group. Engineering I and II were principal subjects in Group II, and their schedules comprised the following papers:—

Engineering I. 1. Practical Mathematics. 2. Applied Mechanics. 3. Properties of Materials. 4. Engineering Physics. 5. Surveying. 6. Essay.

Engineering II. 1. Mechanics of Machines. 2. Theory of Structures. 3. Heat Engines. 4. Applied Electricity. 5. Hydraulics. 6. Mechanical Drawing. 7. Essay.

[2] The lecture list for the Michaelmas Term 1919 contains the following names: Inglis, Ball, Chapple, Dykes, Gordon, Grant, Hayes, Heycock, Knox-Shaw, Lamb, Landon, Lees, Peake, Peel, Portway, Thring, Winter and Womersley (*Reporter*, 1919–20, p. 108). Among the assistant lecturers without university status were H. W. Pear and R. Lubbock.

Lectureship in Thermodynamics was established and endowed with money which Mrs John Hopkinson had raised towards a future Hopkinson Professorship, and on 11 July S. Lees of St John's was appointed to the office at a stipend of £100 a year. From the same date J. W. Landon became University Lecturer in Mechanical Engineering in place of Inglis and W. D. Womersley of Emmanuel succeeded Landon in the demonstratorship. On 10 May 1920, with the consent of the Vice-Chancellor, G. F. C. Gordon and L. C. P. Thring, both M.A.s of Trinity, were appointed respectively superintendents of the Engineering Workshops and Drawing Office. In October 1920, when the undergraduate figures reached their peak, the teaching staff had risen to 48, giving a student–staff ratio of about 17 to 1.

The rapid build up in student numbers after the armistice, which was not, of course, confined to the Department of Engineering, made it necessary to consider as a matter of urgency the provision of new buildings for the scientific departments in which post-war expansion seemed likely to be the greatest. The New Museums site was already overcrowded and both Physics and Chemistry, as well as Engineering, were in desperate need of more space. In the report of 19 May 1919, which has already been quoted, the Syndicate on Engineering Studies expressed the view that existing buildings were incapable of extension and a move should be made as soon as possible to a new site. Scroope House, standing in a garden of over three acres off Trumpington Street, was known to be available and was considered to be admirably suited to the needs of an Engineering School. The matter was referred to the Financial Board which, in a report dated 21 May, recommended the purchase of the site and asked the university's permission to negotiate with Caius College, which owned the freehold, and with their tenant, Dr Wingate. Dr Wingate's lease was not due to expire until 1933, but at the cost of some personal inconvenience he generously agreed to move out if the university and the college came to terms. The sale was completed on Lady Day 1920 for the sum of £14,900.

Inglis did not wait for the completion of the sale before taking action. He secured the appointment of F. W. Troup, F.R.I.B.A., as architect for the new building and during the summer of 1919 plans were prepared and agreed by the department. The original design for the Inglis

Building[1] shows a rectangular building measuring about 300 feet by 210. The Coe Fen frontage was designed to house the workshops and stores and comprised a central two-storey block 100 feet long with single storey wings. The central area had north-light roofing and was subdivided by internal partitions to form three main laboratories, a boiler house and a reading room. The eastern side consisted of an entrance hall with lecture theatres on each side and a second storey containing two drawing offices and two smaller lecture-rooms. Construction was as simple as possible in order to keep costs to a minimum.

On 14 February 1920 a Building Syndicate consisting of the Vice-Chancellor, Inglis, Heycock, Lamb, Landon, Lees and the Building Committee was appointed by Grace and on its recommendation these plans were approved. A tender by W. Saint, Ltd., for levelling the site and putting in the foundations for the first stage of the building was accepted and £2,000 to cover the cost of this was guaranteed by an anonymous donor. The first stage was to consist of the northern part of the main laboratory area and the north wing of the Coe Fen frontage, and the Vice-Chancellor was empowered to sign a contract for this work.

No university money was available to pay for this building, and during the summer of 1920 Inglis formed an Appeal Committee of which the most influential members were Viscount Esher and Mr Dudley Docker. Funds came in slowly, but on 1 October the Vice-Chancellor announced a magnificent offer from Sir Dorabji Tata of Caius College, head of a great Indian engineering combine. The text of his letter was as follows:

Capel House,
New Broad Street,
London E.C. 2.
27 August 1920

Dear Mr Vice-Chancellor,

I gathered, during my recent visits to Cambridge, that the demand for engineering training at the University has so increased in recent years that the existing building and equipment of the Engineering School are now proving quite inadequate; and that you have, therefore, now in hand a scheme for removing, rebuilding, and enlarging the school, which work, if sufficient funds are forthcoming, you expect to complete by October next year. I can quite understand how heavily overtaxed must be the financial resources of the University by the growing

[1] See plan on p. 149.

claims of scientific development. As one interested in a number of engineering and industrial enterprises in another part of the Empire, I fully realise also the value and the need of engineering training at the University in the service of industrial and scientific progress. And I am writing to you to say that, as an old Cambridge man, I shall be only too happy to contribute a sum of twenty-five thousand pounds towards the reconstruction scheme of the University Engineering School. I consider it my privilege to give my old University such assistance as I am able to give for I share the hope that the enlarged school may be the means of imparting a fuller and more thorough training in the subject to the thousands of students who will flock to it from the Empire in future years. I recognise that the University in the past has given a most cordial welcome to young students from India, and given them also of its best. In making this gift, may I venture to express the hope that, in the furtherance of human knowledge, Cambridge will extend to my countrymen a yet warmer welcome; and with the growing demand for higher training in my country, bestow on it the response and the favour of increased facilities for the purpose.

Assuring you of my deepest interest in the success of the scheme,

I remain, dear Mr. Vice-Chancellor,
Yours very faithfully,
D. J. Tata.

Sir Dorabji's offer was accepted, and his gift covered the whole cost of the first stage of the building. W. Saint, Ltd., who had put in the foundations, had already been given the contract for the superstructure, and work continued without interruption throughout the year. Viscount Esher's committee raised about £7,000 by October and it was then decided to complete the whole of Troup's plan except for the eastern block containing the hall, lecture rooms and drawing offices. The university agreed to advance the balance of the money required.

Stage I was completed in December 1920 and the rest of the Coe Fen frontage and the main laboratories early in 1922. The administrative offices were moved to Scroope House in September 1921. The new electrical and wireless laboratories were in full working order when the academic year started in October of that year and the structures, hydraulics and heat laboratories had received most of their apparatus by Christmas, though they were not fully equipped until the end of the following Long Vacation. The old laboratories in Free School Lane were transferred to the Departments of Chemistry and Physics as soon as they were vacated.

The new buildings housed all the laboratories, but in the absence of the two-storey eastern block the areas designed for the workshops had to be used as lecture-rooms and a drawing office. As a result the workshops remained in their old quarters in Free School Lane and some of the lecture-rooms there also had to be retained. The division of the department between the two sites caused inconvenience and loss of time, but nearly ten years were to elapse before the building of the northern part of the eastern frontage made it possible to abandon the old site entirely.

THE FRANCIS MOND PROFESSORSHIP OF AERONAUTICAL ENGINEERING, 1919–1935

On 24 February 1919, when the Professorship of Mechanism was still vacant, the Council of the Senate published a letter from Mr Emile Mond to the Vice-Chancellor offering to endow a Professorship of Aeronautical Engineering in memory of his son Francis, who had been killed while flying on the Western Front. Mr Mond had already obtained a promise from the Air Ministry that if such a chair were established they would station an experimental flight at an aerodrome near Cambridge to work with the university authorities. He stipulated that the university should provide the necessary accommodation and equipment for class and laboratory work within a reasonable time of the appointment of the first holder of the chair. In making this proposal he declared himself to be

> greatly impressed by the advantages which Cambridge offers for instruction and research in a branch of Engineering which seems likely to be of ever-increasing national importance. The nature of the surrounding country, the concentration at the University of distinguished representatives of the Sciences which are closely related to aeronautics, e.g. physics, mathematics and mechanical engineering, and above all the presence at the University of young men of the type of those who naturally take to flying, appear to me to be factors which should be of material assistance in securing valuable results, both from a practical and a scientific point of view.

As his son Francis had been a member of Peterhouse he asked that the Master and Fellows of that College should be represented on the Board of Electors to the Professorship.

Mr Mond's gift was formally accepted on 14 March, and on the following day the Council published a report recommending the establishment of a Francis Mond Professorship 'to promote Aeronautical Engineering in the University by teaching and research'. The electors were to be the Vice-Chancellor, the Master of Peterhouse and eight persons elected by the Senate, and the professorship was to be assigned to the Special Board of Mathematics. The report was approved on 16 May, and on 26 September the election was announced of B. M. Jones, A.F.C., M.A., Fellow of Emmanuel College, who was admitted to the professorship on 1 October 1919.

Bennett Melvill Jones was born on 28 January 1887 at Birkenhead in Cheshire. His father, Benedict Jones, was a graduate of St John's College, Cambridge, where he had taken the Mathematical Tripos before qualifying as a barrister. He had a law practice in Liverpool and was an alderman of Birkenhead, of which town he was mayor in the year 1893–4. He was an enthusiastic and competent amateur engineer, and possessed a well-equipped workshop. The young Melvill went as a day-boy to Birkenhead School where, on reaching the top form, he showed a considerable aptitude for mathematics. It was, however, in his father's workshop that he developed his taste for mechanics and engineering. Together they constructed bicycles, a half-horsepower dynamo and several large storage batteries, which they charged by driving the dynamo from the lathe pedal. Their crowning achievement was a motor car driven by a single-cylinder 4 horsepower petrol engine, which they completed in 1903. It ran successfully for about a thousand miles at a cruising speed of 15 m.p.h.

In 1906 Melvill won an exhibition to Emmanuel College, Cambridge, and in the following year was awarded a scholarship there. He obtained first class honours in the Mechanical Sciences Tripos in 1909. From September 1909 to December 1910 he worked as a 'shop-student' in Woolwich Arsenal and in January 1911 joined the Aerodynamics Department of the National Physical Laboratory, where he was employed mainly on experimental work in wind tunnels. In May 1913 he moved to the London office of Armstrong-Whitworths and worked there on the design of rigid airships until the war started in August 1914. He was then a 2nd Lieutenant in the R.E. Territorials, but about two months after the outbreak of hostilities he was seconded to what is now known as the Royal Aircraft Establishment, where he worked on aircraft armaments. In the spring of 1916 he was transferred to Bertram Hopkinson's new air armament experimental station at Orfordness, and put in charge first of one and later of three of the six departments into which it was divided. He was concerned mainly with the development of gun-sights, bomb-sights, cloud flying and navigation.

Hopkinson, who had recently learned to fly, wished other members of his staff to qualify as pilots and test on active service the apparatus and armaments which they were trying to perfect in their laboratories. At his request Melvill's younger brother B. H. M. Jones, who had recently returned to duty after being severely wounded, was sent with a training aircraft to Orfordness, and under his tuition Melvill and others learned to fly and in due course received their wings. Melvill did a great deal of flying in various types of aircraft in England and France, and early in 1918 he served for five or six weeks as gunner in a Bristol Fighter on the Western Front, with his brother as pilot, for which he was awarded the Air Force Cross. Soon after his return to Orfordness both Hopkinson and B. H. M. Jones were killed in flying accidents and Melvill was transferred to the Air Ministry as deputy to Henry Tizard, where he remained until after the war had ended. In March 1919 he returned to Cambridge as a Fellow of Emmanuel and a member of the staff of the Engineering Department.

When Jones received the news of his election to the professorship he asked the Registrary what arrangements had been made for the establishment of his new department, and was somewhat startled to learn that the university could not provide him with accommodation, money for running expenses or any other facilities. He was, however, given a free hand to get what help he could from the Air Ministry or anyone else who was interested. This appeared to be a breach of the undertaking given when Mond's offer was accepted, but the university was on the verge of bankruptcy and the Registrary evidently—and correctly—believed that by some means or other Professor Jones would produce the results expected from him. Inglis came to the rescue with accommodation and secretarial help and when the Scroope House site was acquired Jones was given a room in the old servants' quarters as his office and part of the cellar for his research work. Later he moved into a wooden hut erected in the garden by the University Air Squadron, and he continued to work there until his retirement in 1952.

For experiments in flight the Air Ministry allocated pilots and aircraft, and many of the special instruments required were borrowed from the Royal Aircraft Establishment. These aircraft were at first used to determine the accuracy with which the sun's altitude could be measured in flight by gravity controlled sextants and to study problems in aerial surveying. In later years they were used extensively to study, in flight, methods for improving the performance of aircraft and their behaviour when controlled by the pilot. In some of these later experiments they were flown by a research student, Mr J. A. G. Haslam, who had been an R.A.F. pilot during the war, and who afterwards became a lecturer in the department.

In June 1920 the university released £200 for the purchase of books of reference on the ground that his stipend had been reduced by this amount on account of the fellowship dividend which he received from Emmanuel. This grant was renewed annually until the improved financial position of the university made more generous allocations possible, and it was used not only for the purchase of books but as a contribution towards the construction of wind tunnels which were made in the engineering workshops.

On taking up his appointment Professor Jones was at once co-opted by the Board of Engineering Studies, and he attended its first meeting on 2 December 1919. At his request a committee was set up to consider how the study and teaching of aeronautics could be related to the other courses in engineering and in May 1920 the Board agreed to recommend that a paper on Aeronautics should be added to Group B subjects in the Mechanical Sciences Tripos. This was approved on 17 May, and the Ordinances were amended accordingly.[1] Lecture courses were arranged, and as a partial offset against the facilities provided by Inglis it was agreed that the fees received for the lectures

[1] The Schedule for Aeronautics was as follows:

'Theoretical and Experimental Aerodynamics and its application to design and the prediction of performance, stability and control of aircraft. The principles of flight and the analysis of performance. Application of the theory of structures and materials to aircraft design. Meteorology and the Physics of the Atmosphere in relation to aeronautics. Theory and practice of methods of navigation suitable for aerial work. Knowledge of maps and the theory of map-making from aircraft. Theory and use of Instruments used in aerial work. Characteristics of aero-engines, not including design.'

on aeronautics should be paid into the Fee Fund of the Engineering Department.

In December 1922 Professor Jones published his first annual report and informed the university of the progress which had been made in establishing his new department. He himself was giving a course of lectures primarily intended for candidates for the 'B' paper on Aeronautics. It was attended mainly by students who hoped to make aeronautical engineering their profession, including R.A.F. officers doing a two-year course in general engineering, but it also attracted a number of second-year men who were interested in the subject without having any particular objective in mind. Five men took the aeronautics paper in the 1921 tripos and the same number in 1922. Thirty-eight men registered for the course in the Michaelmas Term 1922 but most of them ceased to attend as the examinations drew nearer in order to concentrate on the compulsory papers. Lectures on other aspects of aeronautical engineering were also given by Mr W. S. Farren of Trinity[1] (Now Sir William Farren) and on wireless signalling by L. B. Turner of King's. In addition a close liaison was maintained with the work of the Aeronautical Research Committee, of which Jones was a member. One Research Student had studied the failure of light metal structures and one was employed on aerial surveying.

For the next nine years the Department of Aeronautics maintained its independence in theory, though in practice its alliance with the Engineering Department grew continuously closer. In 1927 papers on Aeronautics were included both in Engineering I and Engineering II for the Ordinary Degree and three years later lectures for the tripos candidates on certain aspects of aerodynamics and hydraulics and of the flow of steam in turbines and other engines were combined in a single course under the title of 'Fluid Dynamics'.

By this time it was evident to both Jones and Inglis that aeronautics as taught at Cambridge had become a branch of engineering studies, and with their agreement the Faculty Board of Engineering produced, on 20 October 1931, a report which recommended that the Aeronautics Department should become a sub-department of the Department of

[1] Mr Farren was formally transferred from the Engineering to the Aeronautical Department in 1928.

Engineering, though the Professor of Aeronautics would retain control of his teaching staff and the funds earmarked for aeronautical research. While making this recommendation the Faculty Board also expressed the view that the Engineering Department itself should ultimately be decentralized into sub-departments, each under its own professor. The sub-departments suggested were in Mechanical, Thermodynamic, Civil, Electrical and Aeronautical Engineering, and the amalgamation of aeronautical and general engineering was regarded as a first step towards the ultimate objective. A General Board report dated 1 June 1932 supported the Faculty Board's proposal for the amalgamation of Engineering and Aeronautics and this was approved without opposition on 4 November. The creation of further professorships had to wait until after the Second World War.

The absorption of the Department of Aeronautics with that of Engineering did not impair its prestige, as was proved when on 1 July 1935 Sir John Siddeley gave £10,000, spread over seven years, to help the further development of aeronautical research in the university and to give practical expression to his regard for the work of Professor Melvill Jones and his associates. This donation came at an appropriate moment, as a second wind tunnel had just been added to the equipment of the Aeronautical Laboratory. The first had been designed by Mr W. S. Farren in 1928 and built by F. Gravelin, the laboratory carpenter, and Sergeant May, rigger in the University Air Squadron. The three-component tunnel balance was constructed in the instrument shop by J. Sharpe, A. A. K. Barker[1] and N. B. Surrey.[2] It was in this tunnel that Sir Frank Whittle did some of his research when he was designing his first jet engine. The second tunnel was also designed by Farren and made in the workshops. It is a low turbulence closed return circuit tunnel with a working section thirty-six inches by twenty-eight inches and a maximum air speed of 80 m.p.h. It was installed alongside its predecessor in the wooden hut which the department shared with the University Air Squadron and was officially opened by Sir Henry Tizard in 1936. It was transferred to the new laboratory in the South Wing in 1959 and is still in use at the time of writing.

[1] Now Technical Officer in the Workshops.

[2] Now Chief Technical Assistant in the Aeronautical Laboratory.

CONSOLIDATION, 1923–1939

Seventeen years elapsed between the move of the laboratories to the Scroope House site and the outbreak of the Second World War. During this period the department settled down under new statutes and Inglis raised the standard of teaching to a very high level. Undergraduate numbers, after falling to 407 in 1926 stabilized at a little over 500, rising sharply to a maximum of 582 in the last Michaelmas Term before the Second World War. The number of Research Students working in the department varied between 5 and 21.

After a delay caused by the fall of Lloyd George's government in 1922 and its replacement by the Conservatives under Bonar Law, commissioners were appointed under the Universities of Oxford and Cambridge Act of 1923 with instructions to draw up new statutes for the two universities and their colleges. Viscount Ullswater was elected chairman of the Cambridge Commissioners[1] and H. A. Holland of Trinity was appointed secretary. An excellent liaison was quickly established between the Statutory Commissioners and the Council of the Senate and the great work of revision was completed by October 1925. On the twenty-second of that month a special issue of the *Reporter* carried a complete draft of the 'Proposed Statutes of the University and Statutes for the University and Colleges in Common'. Two days in November were given to their discussion and the amendments made by the commissioners after they had studied the remarks of the speakers were published on 24 November. The Statutes as finally approved by the King in Council were published in the *Reporter* on 29 January 1926. The most important changes which they embodied were the institution of the Regent House[2] to conduct the ordinary business of the university on behalf of the Senate with the help of an administrative staff of university officers, and the establishment of a Faculty system covering all subjects, with a teaching staff of professors,

[1] The commissioners for Cambridge were: Viscount Ullswater; Bishop H. E. Pyle; the Dean of Westminster; Sir Thomas L. Heath, Sc.D, F.R.S.; Sir R. T. Glazebrook, Sc.D., F.R.S.; Sir H. F. Wilson, M.A.; Sir Hugh K. Anderson, M.D., F.R.S., Master of Gonville and Caius; William Spens, M.A., Master of Corpus Christi; Bertha Surtees Philpotts, D.Litt., Mistress of Girton; Peter Giles, D.Litt., Master of Emmanuel; W. R. Rendell, M.A., Fellow of Trinity Hall; E. H. J. N. Dalton, Sc.D., Cassel Reader of Commerce, London University.

[2] Composed of members of the Senate who are resident in Cambridge and working in or for the university.

readers, lecturers and demonstrators and with improved financial arrangements for the control of teaching and research. Professorial fellowships were established and allocated to colleges on a quota system and the regulations governing the contributions of colleges to the University Chest were revised. The exclusion of women from the university and their right only to the title of a degree were confirmed.

On Friday 13 March 1925 the first recorded Conversazione was held in the Engineering Laboratories. The steam engines and other apparatus were run at full power and refreshments were served in the drawing office. A thousand guests were entertained, and among them were distinguished engineers from all parts of the country and senior officers from the Army and Royal Air Force.

The Faculty Board of Engineering, established under the new statutes to replace the Special Board of Engineering Studies, met for the first time on 15 November 1926. The Professors of Mechanism and Aeronautics and the Cavendish Professor were *ex-officio* members, six teaching officers were elected by the Faculty and two more nominated by the Council of the Senate. Two other members could be co-opted.[1] J. F. Cameron was re-elected chairman and J. W. Landon secretary. A degree Committee was also established. It consisted of the same three professors and three other members of the Faculty Board who were elected annually.

The first major task of the Faculty Board was to agree with the General Board on the establishment of a staff of university lecturers for the department. Inglis had presented his views on this matter to the Statutory Commissioners in a memorandum in 1924. The established staff at that time consisted of the Professor of Mechanism, the Reader in Electrical Engineering, the Lecturers in Thermodynamics and Mechanical Engineering, the Superintendents of the Workshops and Drawing Office and three demonstrators—a total of nine. A large part of the instruction was given by twenty-four assistant demonstrators who held no university post and had therefore no security of employment. Inglis suggested that at least half of them should be given

[1] The original members were J. F. Cameron, Professor C. E. Inglis, Professor Sir E. Rutherford, Professor B. Melvill Jones, Dr C. G. Lamb, J. W. Landon, F. J. Dykes, S. Lees, L. B. Turner, J. T. Spittle and A. D. Browne. In the following month R. H. Fowler and Professor Sir J. Larmor were elected to represent the Faculty of Mathematics and C. T. Heycock that of Physics and Chemistry.

university status and that in order to describe the dual nature of their duties they should be known as 'Lecturers and Demonstrators'. This suggestion was not adopted, and the establishment as finally agreed consisted of the professor and reader and twenty-five university lecturers *or* demonstrators.

A scale of stipends for university lecturers and demonstrators proposed by an Initial Appointments Committee in May 1926 was accepted by the General Board in February 1928, and in November of that year a scale was also agreed for university administrative officers, professors, readers and other officials. Faculty Boards were asked to fit their existing staffs into these scales by awarding appropriate seniority, and the Faculty Board of Engineering appointed a standing Committee on Stipends to comply with this request and subsequently to keep a watch on the work done by members of the staff and the payments made to them.

An entirely new system of control over expenditure on teaching and research was established under the new statutes. Faculty Boards were required to present a statement of their needs for the ensuing financial year by 15 November and the total income and expenditure of the University in the preceding and current years was to be published by the Financial Board by 15 December. Recurrent and non-recurrent grants to the various Faculties would then be approved according to the amount of money available and the General Board's assessment of the relative importance of conflicting claims. The first allocations under

Estimates for the year 1926–1927

Receipts	£	Expenditure	£
Fees from students	18,800	Instruments, materials, repairs, apparatus, etc.	3,100
Workshop sales	2,600	Wages, insurance, petty cash	4,800
R.E. students' fees	1,650	Salaries:	
R.A.F. supervision fees	400	From Faculty Fund	11,700
Board of Education course	300	From Imprest Account	3,600
Interest on War Stock	26	Pensions:	
Grant from Chest	2,000	For attendants	450
Stipends and pensions	1,505	For teaching staff	1,190
		Heat, light and water	1,260
		Rates and insurance	500
	£27,281		£26,600

the new system were made in January 1927. The total income for the university for that year was estimated at £195,000 and an expenditure of £191,300 was authorized. Of this the Engineering Department received £2000 in respect of stipends and a non-recurrent grant of £640 towards the purchase of a new testing machine. The estimates for Engineering for the year 1926-27, which are preserved in the Faculty Board Minute Book, show the remarkable self-sufficiency of the department at this time (see p. 166).

In the Michaelmas Term 1926 there were 448 undergraduates and 7 Research Students working in the department.

At about this time the drying out of a ditch which had formerly discharged into the Granta caused a serious settlement of the Coe Fen frontage of Troup's building. Large cracks appeared in the walls near the north-west corner of the Electrical Laboratory and efforts to check further movement by inserting steel ties were unsuccessful. However, in the summer of 1930 a large part of the west wall was underpinned with concrete and further subsidence was prevented. In the same year plans were approved for a new building on the north-east side of the laboratories somewhat similar to that which had appeared in the original design of 1920. The university had just received a benefaction from the Rockefeller Foundation and £25,000 of this was allocated to enable Engineering to complete the move from Free School Lane which had been left unfinished through lack of funds in 1922. A two-storey block containing a lecture theatre for 250 and two lecture-rooms with a students' drawing office above was completed in July 1931 at a cost of £20,100 and when the new academic year opened in October the department was once more united on a single site.

Inglis seized the opportunity of the move to get rid of as much obsolete equipment as possible and reorganize the laboratories. The workshops were moved from Free School Lane to the two-storey block on Coe Fen which had been originally designed for them, and at the same time the temporary drawing office and part of the Electrical Laboratory were converted into lecture-rooms. He was assisted in the general rehabilitation by a gift of several thousand pounds' worth of workshop equipment and scientific apparatus from the laboratory of the late Sir David Salomans, and a further grant of about £2,000 by the university.

While this reorganization was in progress changes were made in the courses and examinations for the Ordinary Degree. It was by this time generally accepted that the papers in Engineering I covered too wide a field for a single Principal Subject and that the engineering examination as a whole was much more difficult than those in other subjects. The syllabus was now spread over three Principal subjects, Engineering I, II and III, and the new regulations were embodied in a report dated 20 May 1927. They did not, however, remain in force for long, as in March 1931 a Syndicate for Courses leading to the Ordinary Degree recommended the abolition of Principal and Subsidiary Subjects and a reversion to the old system of a General Examination and a number of Specials. Engineering, however, was excluded from this scheme and Engineering I, II and III became Examinations in Engineering Studies which qualified for the B.A. degree without the candidate having to take either the General Examination or an arts subject in addition. This was approved on 12 June and for the first time since the original institution of the General and Special Examinations in 1865 it became possible to organize a three-year course devoted entirely to engineering for men who could not cope with the mathematics required for the Mechanical Sciences Tripos. Schedules for the new examinations were approved on 10 March 1933.

In the following year a change was made in the title of Inglis's chair. 'Mechanism and Applied Mechanics' no longer gave a true indication of the field for which the professor was responsible, and in March 1934 the professorship was renamed that of Mechanical Sciences—a term already attached to the tripos and used in the Charter of the Institution of Civil Engineers. An attempt by the Faculty Board to name the chair after Robert Willis, who had done so much to establish the teaching of engineering in the university, was rejected by the Council of the Senate.

No further developments of major importance took place during the five years which immediately preceded the outbreak of the Second World War, but there were numerous changes in the staff. Lamb retired in 1934 from the readership which had been created for him and R. H. Angus was appointed to a demonstratorship which replaced it. Gordon died in 1938 and was succeeded as Superintendent of the Workshops by J. A. D. de Courcy. When Thring reached the retiring age at the end of the same academic year the post of Superintendent of the Drawing Office was allowed to lapse, and his duties were distributed among other members of the teaching staff. R. D. Davies, who was later to replace Landon as secretary of the Faculty Board, joined the department as a demonstrator in October 1934.

The Engineering School at Cambridge was by this time generally recognized as one of the most important in the world by reason of the high standard of its teaching. Inglis made no great innovations in the general scheme of instruction which had been worked out originally by Ewing. He had been closely associated with the reforms brought in by Hopkinson in 1907, when the A and B papers were instituted, and was as convinced as his two predecessors had been that the teaching of engineering should be based on a sound knowledge of mathematics and physics. He defined his views on this matter in his Presidential Address to the Institution of Civil Engineers in 1941.

Premature specialization cramps the imagination and is destructive to the length and breadth of mental vision. Give me a youngster who has had his foundations of belief widely and deeply laid and I will back him, at long odds, to overtake and surpass at his own game one of equal native intelligence who has had his imagination cramped by premature specialization.

In a university course of engineering, instruction should primarily concentrate on teaching those essentials which, if not acquired at that stage, never will be acquired. Technicalities which will automatically be picked up in a student's subsequent career are useful as stimulating interest, but apart from this they are of secondary importance. Education does not consist in the memorization of a number of facts and formulae, useful as these may be when leavened with intelligence. Education at its best should aim at something deeper and more lasting, and the good of education is the power of reasoning, and the habit of mind which remains, when all efforts of memorization have faded into oblivion.

As a lecturer Inglis was in the very first class, and his power of attracting the admiration of his undergraduates was among the greatest of his outstanding qualities. One of his old pupils says of him

He was the most inspired and inspiring teacher that I have ever had, and this surely, must be said by every man who had the supreme privilege of attending any of his lectures. The joy with which he discovered—apparently for the very first time—some vitally important fact in the course of a series of calculations involving two or three blackboards, had to be experienced to be appreciated.[1]

[1] Letter from Sir Edward Herbert to Professor J. F. Baker, 1 July 1952.

Throughout his life he continued his personal research in various fields, but as far as the department was concerned he tended to keep teaching and research in separate categories, and in filling the teaching posts he paid relatively little attention to the research records of the candidates. Indeed the student–staff ratio was so high and the teaching load on the lecturers so heavy that few of them could have spared the time for a sustained course of original research even if they had wished to do so.

His failure to encourage more research in the department may have been partly due to his detestation of administrative work, which increased with advancing years. He was fond of saying that the accomplishments demanded of the head of a large university department, teaching, research and administrative ability, were as widely separated as the corners of an equilateral triangle, that even teaching ability and distinction in research are apt to be in conflict while administrative ability went with neither. He was fortunate in being able to lean heavily throughout the time he was head of the department on his friend J. W. Landon, the secretary of the Faculty Board, and on A. H. Chapman, who after working with Hopkinson during the war had returned to Cambridge at his request and was created Secretary of the Department in 1931.

Inglis did much of his own research at home, working late into the night on the endless mathematical calculations which delighted him. He was appointed in 1923 to the Bridge Stress Committee, of which Sir Alfred Ewing was Chairman, and soon became intensely interested in all the mechanical and structural problems connected with the railways. He made a particular study of the effect of the hammer blow of engines in producing oscillations in girder bridges, and evolved a theory by which the magnitude of the oscillations could be calculated. He was awarded the Telford Medal in 1924 and elected to a Fellowship in the Royal Society in 1930. In the following year he joined an advisory committee set up by the London, Midland and Scottish Railway to encourage the study of scientific problems in connection with railway development and to define those which could best be dealt with extramurally. He followed his bridge stress research by mathematical work on the vibrations of non-uniform beams and of disks, and embarked on a programme of detailed calculations which occupied him to the end of his life. He served on the Councils of the Institute of Naval Architects and of the Institutions of Civil, Structural and Waterworks Engineers. He was chairman of the Cambridge University and Town Waterworks Company from 1928 and was elected president of the British Waterworks Association in 1935. He was also a member of the committee of inquiry into the loss of the airship R 101. Many firms called on his services as a consultant and this work and the contacts it made for him with the engineering industry gave him particular satisfaction.

Other members of the teaching staff had little time for research, but work of outstanding importance was carried out in the department during the inter-war years by G.I. (now Sir Geoffrey) Taylor, Yarrow Research Professor of the Royal Society. He had collaborated for a short time with Bertram Hopkinson on meteorological, navigational and night-bombing problems before the latter's death in 1918, and like Hopkinson and Melvill Jones had learnt to fly in order to try out his theories in practice. Among his first collaborators after the war ended were Miss C. F. Elam (later Mrs Tipper), with whom he worked in London before she moved to Cambridge, W. S. Farren and H. Quinney, who had been Hopkinson's principal research assistant before the war. His work at this time dealt largely with the distortion of crystals and the plastic deformation of metals, and the results were published in a series of papers which have been reprinted in the collected edition of his works edited by C. K. Batchelor.[1] Farren worked mainly in the aeronautics hut which the department shared with the University Air Squadron and Quinney in the old stables of Scroope House. Later R. M. Davies[2] worked with Taylor on the determination of stress–strain and into methods of measuring pressures in explosions, using the Hopkinson bar in a new way. Their joint paper on the mechanical properties of cordite during impact stressing was published in 1942. Another collaborator was Enrico Volterra[3] who worked in the cellar under Scroope House on stress–strain problems in rubber and other plastic materials.

[1] *The Scientific Papers of Sir Geoffrey Ingram Taylor*, 4 vols. (Cambridge University Press, 1958).

[2] Later Professor of Physics at the University College of Wales.

[3] Now a Professor in Austin, Texas.

Throughout the decade before the Second World War the number of undergraduates reading engineering increased with the growth of the university, remaining at approximately 10 per cent of the total undergraduate population. During this time Inglis built up the teaching staff to the authorized maximum of twenty-five lecturers and demonstrators, besides the Superintendent of the Workshops and an assistant in Experimental Research.[1] Some of these men were authorities in their own branches of engineering and a very few succeeded in pursuing their own lines of research, but for most of them undergraduate teaching was their primary occupation, and it was as a teaching institution of outstanding merit that the department under Inglis will be chiefly remembered.

With the outbreak of war in September 1939 the work of the university was disrupted for the second time in Inglis's lifetime. The retiring Vice-Chancellor, E. A. Benians of St John's, in his address on resigning office on 1 October 1940, surveyed the years between the wars.

The outbreak of war in September 1939 brought to a sudden end two decades of activity unexampled in the history of the University. Never before had there been in the same time so rapid an increase in the number of teachers and students, in facilities for research in the arts and sciences or such multiplication of buildings by University and Colleges. The decisive power and leadership restored to the University in 1926 after centuries of subordination was followed by activities impartially distributed round the whole circumference of academic knowledge. Assisted by State funds and private benefactions the University was enabled to build new museums, laboratories and libraries, to take a leading part in promoting pure and experimental science, and to increase the number of its teachers in all the arts subjects, while the Colleges opened wide their doors to welcome a new stream of students. A 'progeny of golden years' was permitted to descend and

[1] The staff of the Department on 1 October 1938 was as follows:—Professors: C. E. Inglis, B. M. Jones.

University Lecturers: R. H. Angus (Sidney Sussex), A. L. Bird, Hopkinson Lecturer in Applied Thermodynamics (Peterhouse), A. D. Browne (Queens'), A. H. Davenport (Sidney Sussex), G. S. Gough (Trinity), R. A. Hayes (Trinity Hall), J. W. Landon, Mechanical Engineering (Clare), R. Lubbock (Peterhouse), A. H. Peake (St John's), H. W. Phear (Caius), D. Portway (St Catharine's), T. R. B. Sanders (Corpus Christi), J. T. Spittle (Pembroke), L. B. Turner (King's), W. D. Womersley (Emmanuel), W. E. Woodward (Trinity), T. C. Wyatt (Christ's).

University Demonstrators: C. R. G. Cozens (King's), P. de K. Dykes (Trinity), N. A. de Bruyne (Trinity), T. R. C. Fox (Jesus), J. Diamond (St John's), B. Cooper (Clare), R. D. Davies (Christ's), A. V. Stephens, Aeronautics (St John's). Superintendent of the Workshops: J. A. G. de Courcy. Assistants in Research: H. Quinney (Fitzwilliam), J. A. G. Haslam, Aeronautics (Corpus Christi). Secretary of the Department: A. H. Chapman (St Catharine's).

bless us. It might have been my privilege to record some further stages in this progress, for the needs of Faculties and Departments continued to grow and new schemes of expansion were before us. But to this activity the war has called a halt. Yet the decades between the two great wars will form a distinctive episode in our history, with its story of rapid growth and of closer contact with the life of the nation.

THE WAR AND RETIREMENT, 1939–1943

After the Munich crisis of 1938 the government invited all civilians with qualifications which might be useful in war to register for national service in an emergency. Many senior members of the university did so, and when war broke out in the following September some joined the armed forces or civil service, and others took up civil defence posts. Professor Melvill Jones was one of the first to leave, as he was called up for national service in May 1939 and put in charge of a new department called the Gunnery Research Unit. About four years later he was transferred to London to act as consultant to the Air Ministry and he remained there until the end of the war. During his absence the sub-department of Aeronautical Engineering closed down, though its apparatus was used by members of Queen Mary College who, as will be described later, were evacuated to Cambridge when the bombing of London started.

Conscription was brought in soon after hostilities started, but men under twenty were at first exempted. Soon, however, part-time military service became compulsory for undergraduates and civil defence duties were assigned to them. Later undergraduates were called up for full military service after four terms, and in 1943 exemption ceased altogether for men reading arts subjects, though engineers, scientists and medical students were allowed to complete their university courses. To maintain the supply of technologists bursaries were offered by the Board of Trade, and at Cambridge the intake of these men by the colleges was only limited by the capacity of the laboratories to receive them. In the last years of the war nearly three hundred bursaries were accepted in engineering, physics with radio, metallurgy and chemistry.

In the Engineering Department, as in the country at large, the period of activity and excitement which marked the outbreak of hostilities was followed by one of suspense and uncertainty, until the fall of France and the threat of invasion called for the greatest effort in our history. It was a source of satisfaction to everyone concerned when the departmental workshops began to make high precision gauges for Woolwich Arsenal. Inglis turned his attention, as he had done in the First World War, to the design of portable bridges, but it soon became clear that his invention had a serious rival in the Bailey Bridge, whose shallow girders were easier to handle than the Inglis tubular components. An immense amount of work was carried out in the departmental workshops in an attempt to adapt the tubular bridge to the army's requirements, but in the end, to Inglis's deep disappointment, the verdict was given in favour of Bailey.

Meanwhile the routine of the department went on much as usual, and although handicapped by shortage of staff lectures and laboratory work were successfully maintained. A considerable upheaval occurred in the summer of 1940 when London University was evacuated, and Queen Mary College arrived in Cambridge as the guest of King's College. From that time until the end of the war the Londoners used the lecture-rooms and laboratories of the Engineering Department, though the separate identity of both teaching staffs and students was maintained. In general the Cambridge men kept to the normal routine of morning lectures and laboratories and evening supervision, while the visitors used the same rooms and apparatus in the afternoons.

In 1941 Inglis's great services to the engineering profession were recognized by his election to the Presidency of the Institution of Civil Engineers. It was unfortunate that his year of office coincided with the crisis which followed the fall of France, when the normal activities of the Institution were inevitably curtailed. His Presidential Address, on the subject of engineering education, was considered one of the best ever presented.

By arrangement with the War Office and Air Ministry courses for candidates for commissions in the Royal Engineers and Royal Air Force were instituted in 1940, and courses for Naval and Royal Marine officers and R.A.S.C. cadets were held from the Easter Term 1943. A University Naval Division was formed in 1942 analogous to the Senior Training Corps and the University Air Squadron. The latter had its headquarters in a wooden hut erected on the south-east corner of the Scroope House Site.

Meanwhile the workshops were becoming more and more occupied with the production of munitions and in research connected with new inventions such as the ASDIC and proximity fuses. A. L. Bird became temporary Superintendent of the Workshops, and A. H. Chapman assumed responsibility for finding the additional craftsmen required and for handling the clerical work which the wartime direction of labour involved. His wife and a number of other married ladies, including Inglis's elder daughter, joined the workshops and were trained in engraving and the use of lathes and milling machines. Of the teaching staff, some spent their free time working as mechanics on the lathes and benches, while others served in the Home Guard. The foreman in charge of the instrument shop was A. A. K. Barker, nephew of A. W. K. Barker who had been the instrument maker for Ewing and Hopkinson. H. Nutting was in charge of the fitting shop, H. N. Bradford of the smithy and R. N. Kemp of the carpenters' shop. F. W. R. Cowles and D. W. Searle were among the senior assistants in the laboratories who combined war work with their normal duties.

Inglis reached retiring age in September 1940, but at the request of the General Board he continued to teach and act as head of the department for another three years, although throughout this period the professorship remained technically vacant. His work for the Institution of Civil Engineers and his personal research occupied most of his time, and he left the day-to-day administration of the department largely in the hands of Landon and Chapman.

It had been decided that teaching posts which fell vacant during the war should be left unfilled, so that men who were absent on national service would not be put to a disadvantage in comparison with those who stayed at home. By the autumn of 1942 sixteen chairs were vacant, sixteen other professors and three readers were absent on national service and less than half the lecturers and demonstrators were at their posts in Cambridge. The scientific departments in particular were handicapped by shortage of staff, and a discussion on the problem of professorial vacancies took place on 17 November. Some speakers were in favour of maintaining the standstill policy in the interests of those on war service, but the contrary view was expressed by Lord Keynes. In a letter to the Vice-Chancellor which was read at the discussion he contended that the policy was unfair both to potential candidates for professorships who were denied the possibility of election and also to existing professors who were compelled to retire but were asked at the same time to continue in full performance of their duties. He proposed that when a vacancy occurred the Electors should choose one of three courses of action. They should either invite the existing occupant to continue in office for one more year or for the duration of the war, or elect in the normal way, or announce that the chair would be left vacant. No formal decision was recorded after this discussion, but Keynes's proposals were adopted in practice and appointments to professorships were made whenever it seemed desirable.

One of the first chairs to be filled was that of the Mechanical Sciences. Inglis's three-year extension as head of the department was due to expire in September 1943 and on 20 March the Vice-Chancellor announced that a new professor would be elected from 1 October. The names of candidates were submitted by 14 May and on 22nd of that month John Fleetwood Baker, Sc.D., of Clare College, was elected.

Inglis did not make a complete break with the department when his successor arrived. He moved to a smaller room in Scroope House and continued to give occasional lectures as long as the war lasted, but he gradually transferred his activities to King's College, of which he was Vice-Provost from 1943 to 1948. He also increased his outside interests, undertaking more consulting work and giving lectures to engineering institutions and schools in many parts of the country. He was awarded the Charles Parsons Medal in 1944 and was knighted a year later. In 1950 he was the principal guest at a dinner given in his honour in the hall of King's College by the Cambridge University Engineers Association which he had founded in 1929 for graduates of the department.[1] It was an occasion which gave him the keenest satisfaction, and those who were present will long remember the speech with which he responded to the toast to his health. On 31 July 1951 he celebrated his Golden Wedding, and the occasion was marked by a message of con-

[1] See Appendix.

gratulation from their Majesties the King and Queen. In the autumn he went as Visiting Professor to the University of Witwatersrand, where he made himself so popular that he was passed on to many other Colleges in the Union. It was an exciting experience, but he was seventy-six and the strain proved too much for him. On his return to England he spent a restful winter in Cornwall, but his strength gradually declined, and the death of his wife at the end of March was followed three weeks later by his own on 19 April 1952.

7 JOHN FLEETWOOD BAKER

Professor of Mechanical Sciences from 1943

THE PRE-CAMBRIDGE YEARS, 1901–1943

John Fleetwood Baker was born at Wallasey, Cheshire, on 19 March 1901. He was educated at Rossall School, just missed service in the First World War and entered Clare College, Cambridge, with an open mathematical scholarship in October 1920. He read engineering under Inglis, his Director of Studies being J. W. Landon, secretary of the Board of Engineering Studies. He satisfied the workshop qualification by vacation work in a shipyard and obtained first class honours in the Mechanical Sciences Tripos in 1923.

Baker's aim was a career in industry. Certainly on going down he had no intention of ever entering the academic world; a Cambridge B.A. seemed, in those days, all that any sensible engineer would ever need. However, work was scarce in 1923 and what there was had to be shared with the returning war veterans. When six months had passed and no job of any promise was forthcoming from the Cambridge University Appointments Board Baker accepted with some misgivings an offer from the Air Ministry of a temporary post to assist Professor A. J. S. Pippard of Cardiff in a structural investigation. The misgivings were soon forgotten. The structural investigation turned out to be the first step in a new airship programme. In 1919 with the disaster to the R 38 over Hull Britain's short-lived exercise in airship construction had officially come to an end, but there were those in the Air Ministry who considered that there was a future for lighter than air craft. An Airship Stressing Panel had been set up by the Aeronautical Research Committee and Baker's duty at Cardiff was to carry out experiments to extend the work of the panel as a first step towards a new design method. The plan for the construction of two new ships was soon announced and Baker transferred to the Royal Airship Works, Cardington, where the government ship, R 101, was to be designed and constructed. Though his time with Pippard had been relatively short it had a profound effect on Baker and resulted in much fruitful collaboration down the years, including a long-lived textbook[1] published in 1936.

The work at Cardington was a remarkable experience for a young engineer. An airship combined all the romance and thrill of an aircraft and a ship. So little experience was available that all design problems had to be approached from first principles and the young designer had to carry real responsibility from the start. Nevertheless, the training element in the experience was rigorous because every decision was examined critically by the whole design team. The heavy design work was enlivened and interspersed by 24-hour long flights in the old ship R 33, which had been reconditioned, on which the design staff acted as observers in the collection of aerodynamic load data. In fact, there could not have been a greater contrast in enthusiasm, excitement and opportunity between this time and that usually spent in those days by the young engineer following a graduate apprenticeship.

The work on airships lasted 4½ years, the last two of which were spent partly at Cardington and partly at University College, Cardiff, where Baker had been appointed an assistant lecturer and where he continued to carry out experimental work on airship structures.

In 1928 he left Cardiff to work for eight years for the steel constructional industry. The competition of reinforced concrete was threatening the use of structural steel and it was felt by engineers that the undoubted qualities of steel as a structural material were not given full rein by the regulations which governed its use.

The steel industry and the Department of Scientific and Industrial Research decided on a collaborative effort to correct the position. A Steel Structures Research Committee was set up and Baker became its technical officer. He was responsible for directing the committee's work and carried out, with his team, all the necessary investigations for the committee other than those on connections which were in the hands of Professor Batho of Birmingham. In 1933 he was invited to the chair of Civil Engineering at Bristol. The S.S.R.C. work was at a critical stage and would have been seriously delayed if its direction had been changed at that time. It was arranged therefore that the S.S.R.C. staff

[1] A. J. S. Pippard and J. F. Baker, *Analysis of Engineering Structure* (1st ed. London, 1936).

should be transferred to Bristol so that Baker was able in August 1933 to take up his work as head of the Department of Civil Engineering in the University and to continue as technical officer to the committee. Thus began six very happy but intensely busy years. The Steel Structures Research Committee work was successful in that complete tests on actual buildings were carried out from which their real behaviour was learnt for the first time and rational Recommendations for Design were published in 1936. However, Baker felt as a result of his eight years study that the elastic behaviour of structures on which the Recommendations, in common with all other methods in the past, had been based could never really provide a satisfactory and economical method of design. He therefore began in 1936 to investigate the behaviour of ductile structures when loaded beyond the elastic range until plastic hinges were formed and collapse occurred. The results he obtained from the experiments he was able to make in three years were to be of immense value during the Second World War and formed the basis of the work which he continued with a large team in Cambridge after the war resulting in a revolutionary method of steel work design which has now been accepted throughout the world.

The work at Bristol was not all research and steelwork design. There were the normal responsibilities of a small but busy teaching department and Baker as Professor of Civil Engineering carried a heavy load of teaching, lecturing to all years on theory of structures and strength of materials together with the related seminar and laboratory work. He also became intensely interested in post-graduate work, attracting students (one from New Zealand) to work for the Ph.D. degree—a rarity in those days in Bristol engineering circles. He also became convinced of the need for post-graduate courses of instruction for men returning from industry as an economical and rapid means of bringing them in touch with the latest advances. He was, in fact, successful in persuading the Senate at Bristol University to allow him to set up a course of advanced instruction in steelwork design. Unfortunately the war intervened before any students were attracted but it remained on the Statutes until 1950.

In May 1939 the Home Secretary formed the Civil Defence Research Committee to advise him on steps which should be taken to protect the country from the effects of air attack. Baker became a member of the committee. As a result, on the outbreak of war he left Bristol and became a Scientific Adviser to the Ministry of Home Security. The massive air attacks that had been forecast did not materialize immediately but Baker was not idle. He found that precautions had been taken only to protect the populace, by the provision of gas masks and air-raid shelters—nothing had been done to minimize the effect of bombing on industrial production. Many key buildings and plants were terribly vulnerable. In particular some of the great factories in which bomber aircraft were being built were so constructed that one well-placed medium-weight bomb would result in the collapse of acres of the factory. He set up a design department in the Ministry of Home Security, evolved methods of minimizing damage, not only to structures but to machines, and began the uphill task of persuading the service and supply ministries to adopt his precautions. He also took a leading part in setting up a nation-wide network of engineers ready to make technical reports on bomb damage when the 'phoney' war came to an end. When it did, in the summer of 1940, all existing air-raid shelters, which had been designed before the war when no one had any real knowledge of the effects of bombing, proved for one reason or another inadequate. Herbert Morrison, who at this time succeeded Sir John Anderson as Minister of Home Security, called upon Baker to deal with the position. This he was able to do with confidence and speed because the reports of bomb damage coming in from the nation-wide network enabled the problem to be posed with precision—the first application of what was to be known later as Operational Research—while the three years' research carried out at Bristol before the war on the collapse of structures proved the basis of a rational method of designing all forms of structural protection. Not only was it found possible to amend most existing shelters so that their efficiency was greatly increased but entirely new designs of much greater efficiency were produced for all types of personnel shelters including the so-called Morrison indoor table shelter of which over one and a quarter million were distributed to householders; all these new designs were officially adopted early in 1941 by the Ministry of Home Security and later by the Service ministries.

For the next three years the exceptional design team of engineers and architects which Baker had collected worked under extreme pressure. It provided the Ministry of Home Security and the local authorities with shelter and other designs to combat new weapons; in addition they were increasingly used by the ministries and local authorities to solve a wide variety of other problems. Not only did this add up to the equivalent of many years of peacetime engineering experience but it convinced Baker by the time he took up his post in Cambridge in 1943 of the tremendous scope there still was for improving engineering production by the application of scientific methods, not only on the material side but in human relations, since the attempts to protect production had brought him face to face with management problems and had shown him the contribution that the social scientist could make.

PREPARATIONS FOR POST-WAR EXPANSION, 1943–1945

Early in August 1943, on his appointment as Professor of Mechanical Sciences and Head of the Department, Professor Baker moved temporarily into residence in his old college, Clare, where he was made a Professorial Fellow. The war had been in progress for nearly four years and Germany controlled the greater part of Europe, but the United States were now our allies and an assault on the occupied territories was confidently expected in 1944. The tide was not yet on the turn, but in the university as in the country at large people believed in the possibility of victory and thought was being given to some of the problems it would bring with it.

Baker spent the remainder of the Long Vacation discussing his plans for the future of the Engineering Department with senior members of the staff and other influential people in the university. When the Faculty Board met on 25 October his initial proposals were ready for presentation. The Board consisted of the Professors of Mechanical Sciences, Aeronautical Engineering and Metallurgy, the Cavendish and Lucasian Professors, representatives of Physics and Chemistry, Mathematics and Fine Arts and six lecturers from the department.[1] Mr H. Thirkill, Master of Clare, was chairman and J. W. Landon secretary.

[1] J. F. Baker, B. Melvill Jones, R. S. Hutton, Sir W. L. Bragg, P. A. M. Dirac, U. R. Evans, P. C. Powell, B. Tomlinson, A. D. Browne, R. D. Davies, R. A. Hayes, J. W. Landon, R. Lubbock, H. W. Phear and T. C. Wyatt. All except Tomlinson were present at this meeting.

The first important item on the agenda after routine business had been disposed of was a draft report to the General Board on the need for more accommodation for research. It clearly foreshadowed a major change in departmental policy, and is worth quoting at length.

The Board have had under consideration the urgent need of accommodation for experimental research in the Engineering Department. The laboratories are large but are so full of machines and other equipment, essential to the teaching of the number of students accommodated, that there is no space for experimental research work other than that which can be carried out on machines and engines at such times as they are not needed for teaching. In addition there are no small rooms or offices available to research workers for writing up their results.

In the past, the old stables and cellars of Scroope House have been used for research, but they are entirely unsuitable for the purpose and are wholly inadequate for the research work at present being carried out in the Department. Next term Professor Baker's research work is to be considerably increased, and this increase is likely to continue.

In the building plans drawn up in 1937 for the completion of the Engineering Laboratory fairly generous accommodation was provided for research as well as additional lecture-rooms. This completion was not possible at the time owing to lack of funds, and since then absolutely necessary lecture-room accommodation has been provided for by the erection of two temporary wooden huts and by using some lecture rooms at Mill Lane. The Board are very strongly of the opinion that, if the Engineering Department is to maintain its present reputation as an Engineering School for teaching and research, and to deal with the large number of Engineering students after the war, funds must be provided for further extensions.

In the meantime, since the requirements for accommodation for research are so urgent and the completion of the permanent buildings impracticable at the present time, the Board propose that a single-bay steel-framed shed, of ground plan 25 feet by 80 feet, be built as soon as possible...Such a building would accommodate much of the new research work now being started by the Professor of Mechanical Sciences and would provide cubicles for the research staff. It would occupy a site originally planned for an Aeronautical Laboratory, but its construction would be such that it could be dismantled and re-erected elsewhere.

It is estimated that the entire cost of such a temporary building at the present time would not exceed £3000...

After a lively discussion the report was accepted and signed by all those present at the meeting. In due course it was approved by the General Board, who added a proposal that a non-recurrent grant should be made for the purchase of research equipment. The recommendations of the report were approved by the Regent House on 11 December, but the site proposed was not accepted by the town planning authority, and it was eventually agreed to demolish the old stables and erect the hut in their place. Work was put in hand during the summer of 1944 and the first 'Structures Research Laboratory' was completed before the end of the year.

It was obvious that the intended increase in research activity would add greatly to the demands for special apparatus, and the next item on the Faculty Board agenda was a proposal to seek permission to fill the office of Superintendent of the Workshops, which had been vacant since the death of de Courcy in December 1940. This was agreed by the meeting and later approved by the General Board, but more than two years elapsed before an appointment was made, and during this period Mr A. L. Bird continued to supervise the workshop administration.

Next on the agenda came a proposal to establish a Professorship of Electrical Engineering, and a committee consisting of Professors Bragg and Baker, Messrs Hayes, Angus and Landon was appointed to look into the matter. The committee favoured the proposal, but before any action could be taken the initiative passed to the Institution of Electrical Engineers, which in May 1944 made a financial offer which will be described in due course.

Finally the Board turned its attention to the requirements of departmental teaching and appointed another committee to consider the regulations for the tripos and Ordinary Degree examinations, the institution of post-graduate courses and the syllabus of courses which had recently been financed by the Institution of Civil Engineers.[1] Apart from a few items of relatively minor importance this concluded the business of the meeting.

Before the Faculty Board met again notice was received of two major inquiries which were being set on foot by the General Board in preparation for the coming of peace. The first concerned the duties and stipends of the university teaching staff and the second the post-war needs of departments for more staff, new buildings and new equipment. The inquiry into stipends extended over many years and led eventually

[1] The committee consisted of Baker, Browne, Davies, Fox, Landon, Phear and Rhoden.

to important changes in the organization of university and college teaching and in the system of payment which will be described in a later section. The inquiry into the future needs of the university was initiated by the General Board in a letter dated 28 October 1943 which asked departments to put forward statements differentiating between their immediate post-war requirements and proposals for long term development. The letter was considered by the Faculty Board on 19 November and the Teaching Committee set up at the previous meeting was asked to study the problem of post-war expansion and produce a draft report for discussion. On 6 January the General Board circulated another letter which indicated that the University Grants Committee was considering post-war development on a national scale, and Faculty Boards were asked for a general policy statement and an estimate of the number of students who could be accommodated in the period of exceptional pressure which was expected when the war ended and later when conditions had become more normal, assuming no additions to existing buildings. The Faculty Board of Engineering replied on 24 January that 750 students might be accepted during the post-war period provided they were fairly evenly divided between first-, second- and third-year courses. Under normal conditions the maximum was stated to be 600, which was 45 more than the total in the Easter Term 1939. The committee's draft report on post-war policy was considered by the Faculty Board on 24 January and a number of amendments were agreed to. The final report was approved at a special meeting on 7 February. It was accepted by the General Board and published later in the year, together with reports by the other scientific departments, in a booklet intended primarily for the use of the University Grants Committee.

An introductory section to the report reminded its readers that when the war broke out the tripos and Ordinary Degree examinations were under review. During the war undergraduate courses in engineering had been reduced to two years, and the consequent adjustments provided useful data for post-war reforms. The Board had now agreed that changes in the examinations and in departmental policy were desirable, and detailed proposals would be made in due course. These intended changes formed the educational background to the statement of needs, which were summarized under four heads. First, the tripos was to be divided into two parts. Part I would cover a general engineering education, and success in it would qualify for an honours degree. Part II would allow greater depth of study in a chosen, more specialized, field. Secondly, the examinations in Engineering Studies were to be completely revised with a view to making them more valuable as a professional qualification. Thirdly, post-graduate courses combining instruction and directed research would be established in certain branches of engineering. Fourthly, research by the teaching staff would be encouraged by reducing the number of hours devoted to formal teaching and by increasing facilities for research students in the laboratories.

The next section was headed 'University Teaching Officers'. It began with a proposal similar to that put forward by Inglis in 1931,[1] that the department should consist of four sub-departments representing the main branches of the profession—Civil, Mechanical, Electrical and Aeronautical Engineering—each under a professor, one of whom would act as head of the department. The report then made a strong case for a large increase in teaching staff. Before the war most of the lecturers had devoted an average of 28 hours a week during full term and the Long Vacation residence to formal teaching, including college supervision. This made it very difficult for them to find time for research, an activity which the Board now considered highly desirable for a teaching officer. To bring the teaching load more into keeping with that of other Faculties the number of lecturers and demonstrators should be raised from 25 to 45. The expected rise in the number of research students would also necessitate a considerable increase in the assistant staff in the laboratories.

The last section of the report made the case for new buildings. Before the war only two members of the teaching staff beside the professors had rooms of their own on the site, and examples were also quoted of gross overcrowding in the laboratories and workshops. The Board asked for additional permanent buildings to provide more lecture-room and drawing office accommodation and new laboratories, the transfer of the workshops to a new building and the erection of an administra-

[1] See p. 163.

tion block on the site of Scroope House. It was estimated that the capital cost of the whole scheme, based on pre-war prices, would be about £273,000, with an increase in annual expenditure of about £15,000. The corresponding figures for the needs which should be given top priority when hostilities ended were £204,000 and £12,500.

The drafting of this report was Landon's last major task as secretary of the Faculty Board. He retired at the end of the academic year and died within a fortnight, on 11 October 1944, at the age of sixty-five. He joined the department as an assistant demonstrator in 1901 and had been secretary since the first meeting of the Board of Engineering Studies on 2 December 1919. Throughout the entire period of Inglis's professorship he had been his devoted friend and collaborator. He was succeeded as secretary of the Faculty Board by Dr R. D. Davies, who had joined the staff as a demonstrator in 1934 and been promoted to lecturer in 1943.

THE BAKER BUILDINGS, 1943–1965

The erection of the first Structures Research Laboratory on the site of the old stables has already been described. The permanent buildings asked for in the 1944 statement of needs had to wait until the government released labour and materials for work on educational establishments in the summer of 1947. However, in June 1944 a syndicate was appointed to consider the building needs of the scientific departments, and in February 1945 the Faculty Board of Engineering asked for the immediate appointment of an architect to prepare plans for the further development of the Scroope House site as soon as this became possible. The Council of the Senate agreed, and on 21 May the commission was accepted by Mr J. Murray Easton, of Easton and Robertson, Chartered Architects, who had done important work for the university before the war.

The first requirement was obviously to complete the east front of the Inglis Building, but instead of putting up a second wing of lecture-rooms as had originally been intended Professor Baker decided to move the workshops into the new wing and convert the space which they then occupied into laboratories and lecture-rooms, and he proposed that a new administration block should be built elsewhere on the site. Dis-

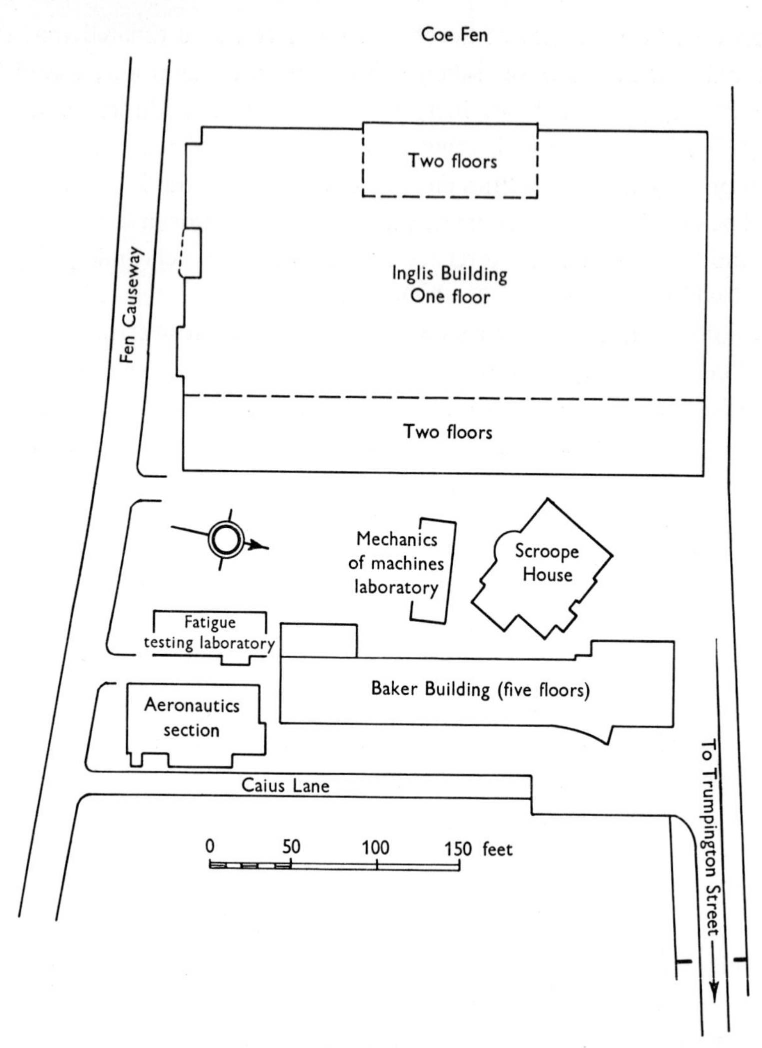

Engineering Department, Cambridge University:
buildings at November 1952.

cussions took place between Mr Easton and the department and a design for a large E-shaped building was gradually evolved. It was to be built in several stages, the first of which would constitute the backbone of the 'E' and contain studies, lecture-rooms, research laboratories and offices for the administration. In April 1946 the Buildings Syndicate announced that an outline development plan for the whole site had been accepted by the department, and they recommended that detailed planning for the completion of the Inglis Building should start at once. The Council approved, designs were drawn up by Mr Easton and tenders were called for. On 17 June plans for a two-storey workshop block were approved by the Regent House, and Messrs Rattee and Kett's tender at £52,391 was accepted shortly afterwards.

There was some delay due to shortage of labour and materials, but work finally started in the autumn of 1946. A water tank which stood on a steel framework and supplied the Hydraulics Laboratory was enclosed in a brick tower and connected to the existing structure and the east front was completed by a two-storey range 132 feet long and 54 feet deep. This contained the machine shop on the ground floor with the instrument and wood-working shops above it; the superintendent's offices and the photographic section were housed in the central block under the water tower. This was the first permanent building to be erected by the university after the war. It was completed in October 1948 and on 10 June 1949 the opening ceremony was performed by the Chancellor, Field Marshal Smuts, who unveiled a mural entitled 'A Short History of Engineering' by Mr Tony Bartl which had been presented to the department by Professor Baker.

When the new workshops came into service the old smithy and fitting shop overlooking Coe Fen were added to the Heat Laboratory as a research area, the old instrument shop was turned into a concrete laboratory and the carpenters' shop on the upper floor of the centre block became a lecture room. At the same time a second underground tank, with a capacity of 28,000 gallons, was constructed under the floor of the Hydraulics Laboratory. Most of this work was completed before the beginning of the Michaelmas Term 1948.

Mr Easton's proposals for the rest of the site were submitted to the University by the Buildings Syndicate in a report dated 20 January 1948. The department's request for increased accommodation had been given high priority by the Council, and with this in mind the Syndicate recommended that the backbone of the 'E' should be built as soon as a permit could be obtained. The Regent House approved the report on 28 February and consultations with the architect started immediately. Plans were completed a year later and details of the new building were given in a report dated 4 May 1949. The contract was let in two parts, for foundations and site works to Rattee and Kett, Ltd., at about £30,000 and to W. Sindall, Ltd., at £195,000 for the superstructure. The kitchen wing of Scroope House and the temporary Structures Research Laboratory were demolished during the Long Vacation to make room for the new building and work started on the foundations. The building was taken over in July 1952 and the opening ceremony was performed by the Duke of Edinburgh on 13 November in the presence of the Chancellor, Lord Tedder, members of the department and two hundred and fifty guests.

The main block of Baker Building is 256 feet long, with a total floor area of about 54,500 square feet. It has five main storeys, plus a mezzanine and a basement, and contains about fifty studies for the teaching staff and Research Students, lecture-rooms, laboratories, two Common Rooms, the Board Room, the departmental library and offices for the administration. It made possible many of the developments in teaching and research forecast in the 1944 statement of needs. In particular the new studies, together with rooms in Scroope House freed by the removal of the administration and library, enabled every member of the teaching staff to have his own room, while the Common Rooms proved invaluable for the exchange of information and ideas and the entertainment of visitors. These facilities were particularly important for the growing number of teaching officers who did not hold college fellowships.

It was not long, however, before the new accommodation proved inadequate to meet the growing needs of the department. On 1 October 1952 W. A. Mair succeeded Sir Melvill Jones as Professor of Aeronautical Engineering. He found that the wooden hut which had formerly belonged to the University Air Squadron was filled to capacity with three small wind tunnels and other research apparatus. He wanted space for a much larger wind tunnel to produce an air stream of low turbulence and a supersonic tunnel as well as laboratory space for other research. He was also convinced of the need to relate the study of fluid mechanics to other branches of engineering besides aeronautics, and in conjunction with the newly appointed Professor of Applied Thermodynamics produced a scheme for putting all fluid mechanics and heat transfer work into adjacent laboratories in a block which might form the southern wing of Mr Easton's 'E' shaped Baker Building. As this would encroach on the site of the existing Aeronautics and Fatigue huts he also suggested that a single-storey laboratory should be built elsewhere on the site to provide a temporary home for his own equipment and a permanent Fatigue Laboratory. His paper, which was entitled 'Notes on a Proposed New Fluid Mechanics Laboratory for Mechanical and Aeronautical Engineering' was presented to the Faculty Board on 26 February 1953 and passed to the Secretary General for information, though it was known that no money was immediately available to implement these proposals.

However, in the summer of 1953 the government decided to assist the further development of technological education by giving increased grants to certain universities in which schools of engineering and applied science were already well established. The Vice-Chancellor received a letter from the secretary of the University Grants Committee in August asking for proposals from Cambridge, and the General Board invited the Faculty Boards concerned to amend their current statements of needs in the light of the new policy of technological expansion. Professor Baker submitted his proposals on 28 September, asking for more teaching and assistant posts and funds for the addition of two wings to Baker Building. When the reports of the other Faculties had been considered the General Board published a report, dated 3 November 1954, on developments in higher technological education which outlined the government's proposals and recommended their acceptance both in the national interest and in that of the university as a whole. It was well known that some members of the Senate were apprehensive lest the expansion of technology should upset the balance of studies at Cambridge, but the Council had taken precautions which were generally considered adequate and the report was approved without opposition on 6 December.

A further report by the Council published on 23 May 1955 dealt solely with the extension of the Engineering buildings, and proposed that two four-storey wings with a single storey laboratory between them should be added to Baker Building. The U.G.C. had agreed to meet the entire cost of the buildings and make a grant towards their equipment. The report was approved on 24 June, and to clear the site the Mechanics of Machines hut was removed during the Long Vacation by the local Boy Scouts, who were allowed to keep the hut in return for their labour. The contract for the centre wing and single storey laboratory, known as Stage IIA, was awarded to Coulson and Son, Ltd., at a sum of £71,319 and work started in October.

The steel work of Stage IIA is of particular interest as it was designed by the plastic method by Professor Baker and his team, Dr M. R. Horne and Dr J. Heyman. The main beam-to-stanchion connections were welded on site, seating brackets being provided on the stanchions and special clamps holding the beams in contact with the flanges during welding. Butt welds between the beam flanges and the stanchion and fillet welds on either side of the beam web enabled the joints to transmit the full plastic moment of the beam. This method of construction resulted in a saving of weight of 20 per cent for the main members and enabled beam depths in the centre wing to be reduced to 14 inches instead of the 20 inches which would have been required in the orthodox method of construction.

The single storey laboratory was completed early in the Michaelmas Term 1956. The southern bay replaced the wooden hut as a Mechanics Laboratory and the remainder housed the old wind tunnels until the south wing was completed. The centre wing was taken over at the end of the following Long Vacation. Most of the ground floor became a Fatigue Laboratory in place of the old wooden hut and Structures Research expanded into the remainder. The first floor was allocated to the rapidly growing Control Engineering group and the two upper floors contained a small lecture room and twenty studies. The total floor area of IIA was about 14,750 square feet.

Detailed drawings for the south wing, known as IIB, were completed early in 1956, and the building contract was awarded to W. Sindall, Ltd., at a sum of just under £150,000. Professor Mair's suggestion that other branches of fluid mechanics should be accommodated was adopted, and the first floor was given to the Heat Group as a thermodynamics research area. Aeronautics took the two top floors for its wind tunnels, and the ground floor was divided between lecture and seminar rooms and an open area which was used as a sort of transit camp for research workers during subsequent building operations. The basement houses store rooms and air compressors and air bottles for the supersonic wind tunnel. A goods lift serves all floors. The total floor area of IIB is about 28,700 square feet. It was completed in September 1958.

The growing realization of the vital importance to the national economy of an improved system of technological education led to a general expansion of university engineering schools throughout the country. At Cambridge the number of undergraduate and research students in the department continued to rise, and in 1956, while Stage II was still in progress, Professor Baker decided to press for a further grant from the U.G.C. to complete the 'E' shaped building by the erection of its north wing. Permission to start planning was given by the General Board, and in July 1958 the Financial Board agreed to include the new building in the programme for 1963 at a sum which was not to exceed a quarter of a million pounds. As the work would involve the demolition of Scroope House temporary accommodation was allocated to the department in two of the Scroope Terrace houses adjoining the site. Plans were approved by the city authorities in August 1961 and in December the building contract was awarded to Kerridge (Cambridge), Ltd., at a sum of £201,702. Work started on 28 January 1963 and was completed early in 1965.

The structure of the north wing, like that of the centre one, is of considerable technical interest as it is designed on a further development of the plastic theory. Research on the behaviour of composite steel and concrete structures started in the department in 1961 under Mr R. P. Johnson, who had been appointed to a lectureship in the Concrete Group after working for four years on the staff of Ove Arup and Partners, Consulting Engineers. The north wing provided an excellent opportunity for trying out the first results of Mr Johnson's research, as it was a simple five-storey building with single spans of 48 feet. Seven different designs were produced, four by Mr Johnson and Dr Heyman on the composite plastic theory and three by the Consulting Engineers, Messrs R. T. James and Partners, who co-operated in the investigations and priced all the proposed solutions. One of the plastic-composite designs, with the concrete floor slabs connected by welded studs to a rigid-jointed frame, proved to be the cheapest and gave a floor construction only 25 inches thick—6 inches less than the corresponding figure for a traditional steel structure. The resulting increase in headroom was peculiarly valuable as floor levels were determined by those existing in the adjacent main block, whose spans were much smaller. This design was therefore adopted, and the north wing is believed to be the first plastic-composite structure to be erected in this country.

The most important feature of the north wing is a lecture theatre to seat 334 people, with facilities for 35 and 16 millimetre projection, closed circuit television and simultaneous translation. In order to provide maximum floor space above it within the building height allowed by local bye-laws the ceiling of the lecture theatre is level with the mezzanine floor of the main building, and the seating slopes down to the floor of the lecturer's dais fourteen feet below ground level. Rooms for a professor, nine lecturers and an enlarged Central Registry are provided on the mezzanine floor. On the first floor a new drawing office bridges the central road and joins up with the drawing office in Inglis Building. The original Soil Mechanics Laboratory on the second floor has been extended twenty feet into the new wing and the constant temperature room doubled in area. The second floor replaces the former Part II Electrical Laboratory in the Inglis Building and the top floor is given over to electronics research. Between the north and centre wings the ground has been excavated to form a storage basement with an area of over 5000 square feet and a paved court above it. An external lift in a glass-sided tower serves all floors and the basement. The floor area of the north wing, excluding the basement, is nearly 23,800 square feet. This brought the total floor area on the site, excluding basements, to about 200,000 square feet.

The two top floors of the north wing were completed and handed over by the contractors in September 1964. Teaching for the Electrical Sciences Tripos[1] was transferred from its old quarters in Inglis Building to the second floor during the Michaelmas Term and the research laboratories on the third floor were equipped and put into use early in 1965. The new lecture theatre was, by a great last-minute effort, finished in time for the Duke of Edinburgh to open a conference there on

[1] See p. 225.

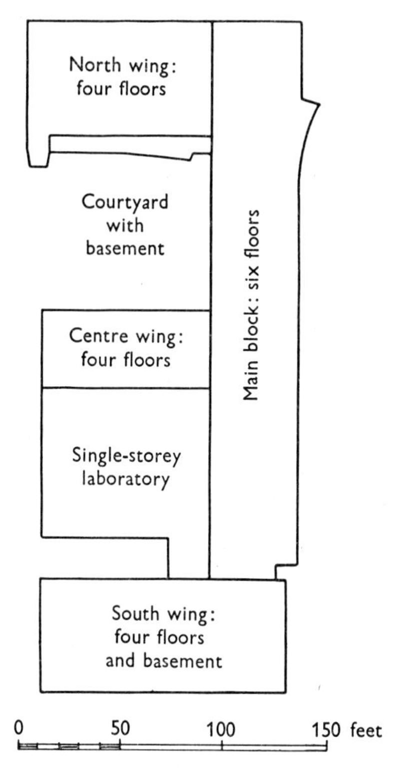

Baker Building, completed 1965. Architects: Easton, Robertson, Cusdin, Preston and Smith.

Key to Plan of Baker Building, 1965.

MAIN BLOCK. *Basement*, Stores, Lavatories, etc. *Ground floor*, Lecture-room, Structures Research Laboratory. *Mezzanine*, Staff Rooms and Offices. *First Floor*, Library, Lecture-rooms. *Second Floor*, Board Room, Common Rooms, Staff Rooms, Soil Mechanics Laboratory. *Third Floor*, Staff Rooms. *Fourth Floor*, Research Laboratories, Staff Rooms.

NORTH WING. *Ground Floor*, Lecture Theatre. *Mezzanine*, Staff Rooms. *First Floor*, Students' Drawing Office. *Second Floor*, Electrical Sciences Tripos Laboratory. *Third Floor*, Electronics Research Laboratory.

CENTRE WING. *Ground Floor*, Structures Research, Fatigue, Mechanics Laboratory. *First Floor*, Control Engineering Laboratory. *Second and Third Floors*, Staff Rooms.

SOUTH WING. *Basement*, Stores, Compressor Room. *Ground Floor*, Lecture-rooms. Mechanics, Metrology Laboratories. *First Floor*, Thermodynamics Laboratory. *Second and Third Floors*, Aeronautics Laboratory.

30 March. The basement stores under the central court were in use by midsummer.

The north wing completed the E-shaped building which the university had approved in February 1948. At that time it had seemed that the new accommodation would meet the needs of the department for the next half century, but so rapid had been the expansion of technological education and so effective the co-operation of the university authorities and the University Grants Committee that the whole scheme had been carried out in seventeen years.

The north wing was officially opened by Lord Nelson of Stafford, president of the Cambridge University Engineers Association,[1] at a ceremony in the new lecture theatre on 3 December. It was preceded by a lunch in the Board Room at which the Vice-Chancellor, Mr A. Ll. Armitage of Queens', presided. Guests included the architect of the later stages of the building, Mr S. E. T. Cusdin, past and present members of the Building Committee and representatives of the four firms of contractors responsible for the buildings. The opening ceremony was attended by many senior members of the university, including heads of Houses and professors, the graduate and senior assistant staff of the department and distinguished representatives from industry and other universities. In the evening there was a conversazione at which all the laboratories were open to invited guests and the work of over one hundred of the research students was demonstrated. A similar conversazione had been held on the previous evening, the last day of full term, for undergraduates, the assistant staff and their friends. About two and a half thousand people attended these gatherings.

The height of the north wing as originally designed had had to be reduced to meet the requirements of the city planning authority, and it was at first intended to replace the lost floor space by rebuilding part of the Coe Fen frontage of the Inglis Building as a multi-storey block. Further examination, however, showed that this was undesirable and in December 1961 Mr S. E. T. Cusdin[2] was asked by the Engineering Extension Building Committee to review the whole problem of site

[1] See Appendix, p. 260.

[2] Then of Easton, Robertson, Cusdin, Preston and Smith. Now of Cusdin, Burden and Howitt.

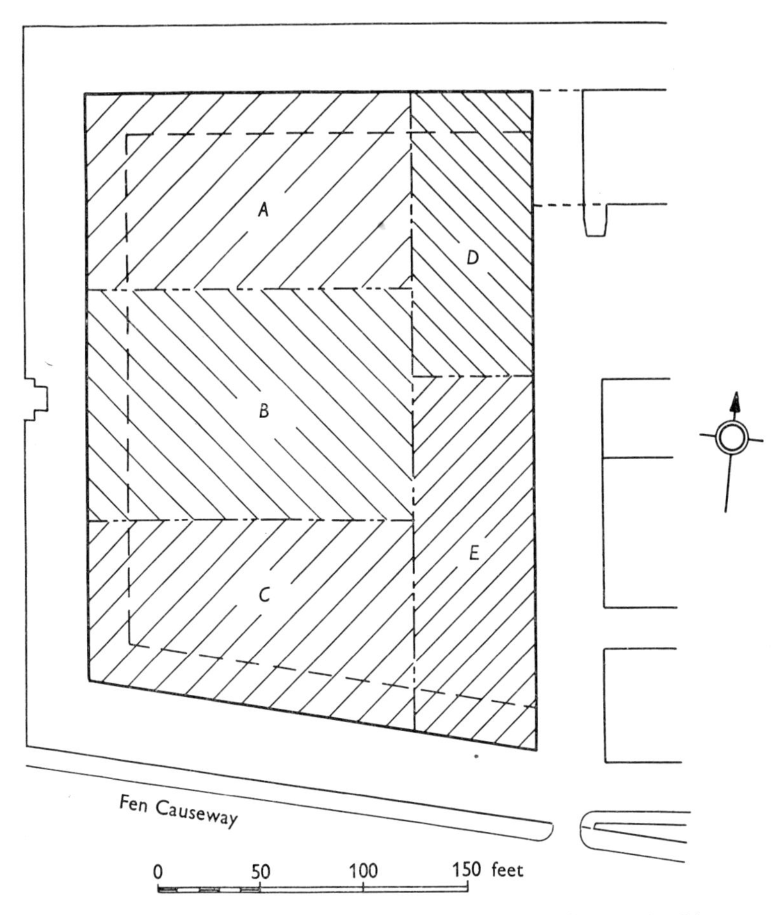

Plan for the rebuilding of the Inglis block, approved 1963. Architects Easton, Robertson, Cusdin, Preston and Smith.

development and make new proposals. These were accepted by the Building Committee and submitted to the university by the Council of the Senate in a report dated 28 May 1962 which was approved on 29 June. The plan involves the rebuilding of the whole Inglis Block, including the 1931 lecture theatres and the 1948 workshops, in five stages, which will be known as Inglis A to E. The final building will be nearly rectangular in shape, three storeys high with a maximum height of 48 feet, surrounded by a ring road and set back on three sides on the ground floor to allow undercover car parking. It is designed as far as possible on the open plan principle and the outer walls will be of glass.

A further report dated 22 July 1963 proposed the inclusion of Inglis A in the major building programme for the following year, and this was approved on 27 November. The demolition of the old electrical laboratories began in November 1964. The Part I Teaching Laboratory was moved to temporary quarters in the machine shop of the old University Press building in Silver Street during the Long Vacation. Most of the research equipment was transferred to the top floor of the north wing and the remainder to the Press site, where it will remain until the new buildings are completed in November 1966. The building contract has been awarded to Rattee and Kett, Ltd., at a sum of £351,877.

Inglis A will replace the electrical laboratories built in 1921 and will occupy a ground area, excluding the car bays, of about 10,000 square feet. A mezzanine floor will have a void in the centre, about 40 feet square, to enable tall equipment to be placed on the ground floor below. The first floor will be an open laboratory and the second, which will be set back 16 feet from the Coe Fen frontage, will have studies for research students on the northern side. The ground and mezzanine floors have been allocated to structures and materials teaching and research, the first floor will form the Part I Electrical Teaching Laboratory and the second, which will be top lit, will be given to electrical research. The total floor area of Inglis A will be about 40,800 square feet, giving an additional floor area of 26,000 square feet and parking for twenty more cars.

As in the case of the north wing the design of the steel frame of Inglis A was given to Dr Heyman and Mr Johnson. By this time the Institution of Structural Engineers and the Institute of Welding had produced a Joint Committee Report on the design of full rigid multi-storey steel frames. Dr Heyman and Professor Horne had been members of the committee, and the report gives what is probably the first code of practice for the plastic design of steelwork. The method approved by the committee was used for the design of the frames for Inglis A and has produced an economical site-welded structure.

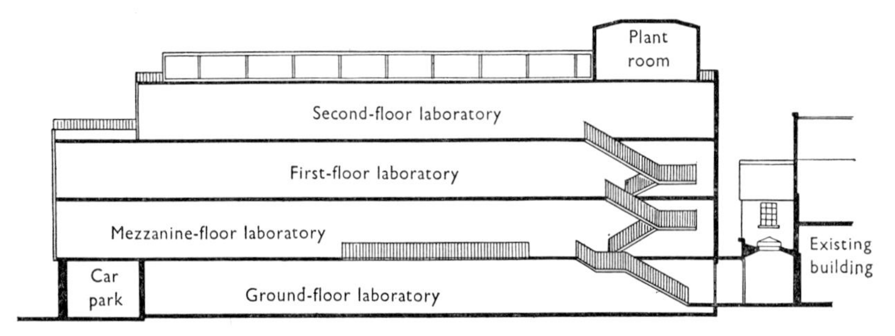

Section elevation of Inglis stage A

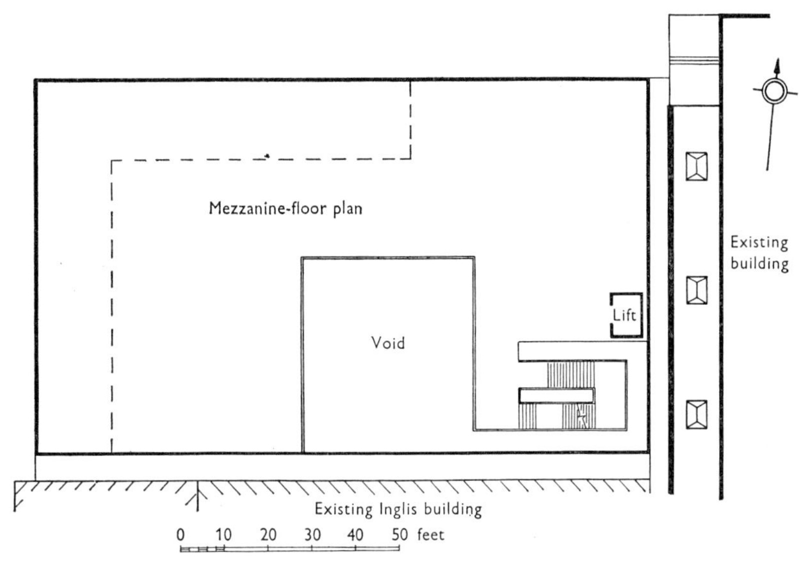

Stage A of Inglis rebuilding plan. Building started in 1964.

THE EXPANSION OF THE DEPARTMENT, 1945–1965

The professorships

When Aeronautical Engineering joined the general Engineering department in 1931 the Faculty Board had suggested that this should be the first step towards the establishment of sub-departments in the five main branches of the profession, each with its own chair, but with one of the professors acting as head of the department as a whole.[1] A somewhat similar suggestion was made in the 1944 statement of post-war needs,

[1] See p. 163.

but by this time the term 'sub-department' had acquired a special significance in the university, and this made it unsuitable for use in a department whose unity both Professor Baker and the Faculty Board were determined to maintain. Later, when several new professorships were established, the teaching staff was organized in groups, each with a professor, reader or lecturer as chairman. The first step in this direction was taken in October 1943 when Professor Baker's first Faculty Board meeting suggested the establishment of a chair of Electrical Engineering. This post was given first priority in the 1944 statement. No university money was available at that time, but on 29 May 1944 the Council of the Senate published a letter from the secretary of the Institution of Electrical Engineers offering to pay the professor's stipend for a limited period while arrangements were being made for a permanent endowment. The text of the letter, which was dated 15 May 1944, was as follows:

Dear Mr Vice-Chancellor,

For many years the Council of this Institution have keenly desired that there should be established in the Engineering School of the University of Cambridge a Professorship of Electrical Engineering, and they have known that the founding of such a Chair has been contemplated in the past.

It may well be, in considering plans for post-war reconstruction, that this matter has been re-opened and that ways and means of carrying it out are being considered. The President and Council have therefore directed me to inform you of their desire to assist the University in such circumstances, so that the Professorship may become effective at the earliest possible moment. To this end the Council are prepared to offer the University a sum of money which it is believed should provide the stipend of a Professor for a limited period of, say, five years, in order that the Chair might be established without delay pending the completion of arrangements for a more perpetual endowment.

During the discussion on this proposal, which was very enthusiastically endorsed at a recent meeting of the Council, there were indications that the Electrical Industry would offer to support the Council's contemplated action by complementary assistance in kind by providing additional equipment for the Electrical Laboratories in the Engineering School.

I am therefore to ask you, Mr Vice-Chancellor, whether your University would be prepared to discuss the matter in more detail with a view to the immediate transfer of the necessary funds, so that there may be no delay in the inauguration of the plan.

Yours faithfully,
W. K. Brasher

Although the Institution's offer made no provision for an endowment the Council of the Senate considered the immediate establishment of the chair important enough to warrant a departure from normal practice. They recommended that the offer should be accepted and that the university should assume responsibility for making further provision for the professorship if no permanent endowment could be obtained from other sources. This proposal was approved by the Regent House on 4 November, and a Board of Electors was nominated. On 28 April 1945 they announced the election of Eric Balliol Moullin, Sc.D., of King's and Downing Colleges, Fellow of Magdalen College, Oxford and Donald Pollock Reader in Engineering Science in the University of Oxford.

Eric Moullin came of an ancient Guernsey family which had seigneurial rights in the island. He was born on 10 August 1893, and being a delicate child was educated privately at home. In 1912 he won an open scholarship to Downing College and the Pooley Scholarship awarded by the Draper's Company. After taking second class honours in Part I of the Mathematical Tripos he obtained first class honours in the Mechanical Sciences Tripos in 1916. He was elected to a Senior Research Fellowship in the Manchester School of Technology, but after a year joined the staff of the Royal Naval College, Dartmouth. While serving there he won the John Winbolt Prize (open only to Cambridge B.A.s) for an essay on 'Some problems of Gaseous Explosions'. When the war ended he returned to Cambridge as an assistant lecturer in the Engineering Department under Inglis and worked mainly in Lamb's electrical group. He also acted as Director of Studies to the officers of the Royal Corps of Signals, and in 1929 was made an assistant lecturer in King's College. He was promoted to university lecturer under the 1926 Statutes.

By this time he had made a considerable name for himself in research. He was awarded research premiums by the Wireless Section of the Institution of Electrical Engineers in 1923 and the two following years and became a member of the council of that body in 1927. In 1930 he moved to Oxford as Donald Pollock Reader under Professor Sir Richard Southwell. He was a member of the Radio Research Board of the Ministry of Scientific and Industrial Research from 1934 to 1942 and chairman of the Radio Section of the Institution of Electrical Engineers in 1939. He also carried out valuable work on vibrations in ships and invented the Moullin torsion meter and voltmeter and numerous electrical measuring instruments. During the Second World War he served first as a temporary Senior Experimental Officer at the Admiralty Signals Establishment at Portsmouth and later on the senior research staff of the Metropolitan-Vickers Electrical Company at Manchester. In both posts he applied his outstanding mathematical and experimental abilities to the development of radio and radar devices. While holding the Cambridge professorship he continued to take an active part in the work of the Institution of Electrical Engineers, of which he was President in 1949–50. He was the author of numerous books and scientific papers on aerial systems and electrical machines.

A permanent endowment for the Professorship of Electrical Engineering was not the only requirement to meet the programme of post-war expansion summarized in the 1944 statement of needs.[1] Money was also required for the endowment of a chair in Applied Thermodynamics, for the erection and equipment of new buildings and for an increase in the staff of university lecturers and assistants. Other departments had also submitted proposals for expansion and it was clear that in the chaotic state of the post-war economy the Treasury would only be able to provide a fraction of the money needed. Professor Baker decided therefore that, in the case of engineering, an appeal should be made to industry, and after negotiation with the university authorities he secured the agreement of the Council of the Senate to his proposals. A 'Note on the Needs of the Engineering Department, Cambridge University' was approved on 30 September 1948. It gave a brief description of the undergraduate courses available in the department and the changes in the curriculum which had been brought in since the war. The need was stressed for a permanent endowment for the chair of Electrical Engineering, the establishment of a Professorship of Applied Thermodynamics, and a substantial increase in the number of lecturers and university assistants as well as for new buildings, new equipment and higher recurrent expenditure on supplies for teaching

[1] See p. 185.

and research. An official appeal for funds to achieve these objectives was launched in the following December.

In certain quarters the reaction to the appeal was hostile. The critics contended that the financing of technological education was a government obligation, that from the point of view of industry the time was inopportune as firms were fully occupied with their own post-war problems and that the failure of the appeal, which was confidently predicted, would make it impossible to try again when conditions became more favourable.

Professor Baker, however, was neither convinced nor deterred. He wrote personal letters to his many contacts in industry, to members of the C.U.E.A. who were in key positions[1] and to anyone else who seemed likely to be able and willing to help. As a result the critics and pessimists were routed and the appeal was a triumphant success. Donations flowed in from many quarters and by September 1950 more than £100,000 had been received. In March 1951 the British Electrical and Allied Manufacturers' Association presented the Vice-Chancellor with a cheque for £71,000 which, with about £2,000 from other sources, completed the endowment of the Professorship of Electrical Engineering. Imperial Chemical Industries Ltd. made seven annual payments of over £7,000 each under covenant to raise the Hopkinson Lectureship to a Hopkinson and Imperial Chemical Industries Professorship in Applied Thermodynamics. Lists of donations were printed annually in the Departmental Report, and by 1959, when all the covenants had matured, the total collected stood at £168,870. As a result, the expansion of the department was able to get under way before the stabilization of the national finances enabled the U.G.C. to implement the policy of improved technological education which will be described later in this chapter.

In 1945 the Shell Group of Companies endowed a Chair of Chemical Engineering in the University. Much discussion took place as to whether the new professorship should be attached to the Department of Chemistry or that of Engineering, but it was finally decided that Chemical Engineering should stand by itself, unattached to any Faculty.

[1] See Appendix.

However, a close liaison is maintained between the three departments, and the Professor of Chemical Engineering is a member of the Faculty Boards of both Engineering and Physics and Chemistry. The first Shell Professor of Chemical Engineering was T. R. C. Fox, a lecturer in the Engineering Department who, as an undergraduate, had achieved the unique distinction of winning the Rex Moir Prize for the best performance in the Mechanical Sciences Tripos and the three other prizes for the best performance in Thermodynamics, Aeronautical Engineering and Structures in the same examination. Professor Fox resigned the chair due to ill health in 1959 and returned to the Engineering Department as a university lecturer. He died on 5 October 1962 at the age of 50. He was succeeded in the chair by Professor P. V. Danckwerts.

The foundation of a Professorship of Thermodynamics was first proposed by Mrs John Hopkinson in 1919, when she and her family contributed generously to the endowment of a university lectureship in that subject.[1] The professorship formed part of Inglis's plan for decentralizing the department and was given second place to the chair of Electrical Engineering in the 1944 statement of post-war needs. It was, as described above, the second object of the appeal in 1948, and in this case the chief supporter was Imperial Chemical Industries, Ltd., which gave £50,000 in seven annual instalments towards an endowment. With this and the money already raised by Mrs Hopkinson a Hopkinson and Imperial Chemical Industries Professorship of Applied Thermodynamics was established from 1 October 1950, and the election of W. R. Hawthorne, B.A., of Trinity College, Cambridge, was announced on 20 September 1951.

William Rede Hawthorne, son of William Hawthorne, M.Inst.C.E., was born at Benton, Newcastle-on-Tyne, on 22 May 1913. He was educated at Westminster School and Trinity College, Cambridge, where he read Mechanical Sciences. He matriculated in 1931 and during the vacations in his first two years did survey work on various sites in England and Scotland. In the Long Vacation of 1933 he worked as a student apprentice with the English Electric Company at Stafford. In the Tripos Examination in 1934 he won the Rex Moir and Ricardo

[1] See p. 155.

Prizes and obtained first class honours with distinction in Applied Mechanics and Heat and Heat Engines. From 1935 to 1936 he was a graduate apprentice in the workshops of Babcock and Wilcox, Ltd., at Renfrew in Scotland and then spent two years as a Commonwealth Fund Fellow in the Chemical Engineering Department at the Massachusetts Institute of Technology, where he received the degree of Sc.D., for research in laminar and turbulent flames. Returning to Babcock and Wilcox in 1939 he worked on problems of heat transfer, measurement of gas temperatures in boiler furnaces and radiation in furnaces, and was put in charge of experimental work on forced circulation boilers. Early in 1940 he joined the Royal Aircraft Establishment, Farnborough, and worked on problems of heat transfer in radiators, oil coolers and de-icing. From the R.A.E. he was seconded for a year to Power Jets, Ltd., at Lutterworth, where he worked under Sir Frank Whittle on combustion chamber development for the jet engine. In the summer of 1941 he returned to Farnborough as head of the newly formed Gas Turbine Division and became responsible for all government research and development work on gas turbines and jet propulsion and for the construction and equipment of the first phase of the laboratories which now constitute the National Gas Turbine Establishment. In June 1944 the work and facilities of the Gas Turbine Division were transferred to Power Jets, Ltd., and many of the R.A.E. staff went with them. Dr Hawthorne, however, was sent to Washington to work with the British Air Commission. He returned to England in 1945 to become deputy director of Engine Research in the Ministry of Supply, but in 1946 he left to join the staff of the Massachusetts Institute of Technology. He took part in the building of their Gas Turbine Laboratory and originated courses in jet propulsion and gas turbines. In 1948 he was elected George Westinghouse Professor of Mechanical Engineering. He assumed his duties as Professor of Applied Thermodynamics at Cambridge on 1 June 1951.

Before the establishment of the fifth professorship two changes took place in the holders of existing chairs. The first was in the Professorship of Aeronautical Engineering. Sir Melvill Jones retired on age in 1952 and was succeeded by W. A. Mair, M.A. of Clare College, on 1 October.

William Austyn Mair was born on 24 February 1917. He was educated at Highgate School, from which he won a scholarship to Clare College. He obtained first class honours in the Mechanical Sciences Tripos in 1939 and then spent two years as an Engineering Pupil at the Aero Engine Experimental Department of Rolls Royce, Ltd., at Derby. In January 1940, after the outbreak of war, he joined the technical branch of the Royal Air Force. He was posted to the Royal Aircraft Establishment, Farnborough, and worked in the Aerodynamics Department there until July 1946. He was demobilized in the preceding February, having reached the rank of squadron leader. At Farnborough he took part in some of the earliest work on the investigation of problems of aerodynamics at speeds approaching the speed of sound and on the later stages of the design of the R.A.E. high-speed wind tunnel which was completed in 1942. In 1946 Manchester University decided to establish a Fluid Motion Laboratory, and Mr Mair was appointed reader and director of the laboratory. After somewhat difficult negotiations he obtained the lease of a hangar at Barton Airport outside Manchester and there, with the help of surplus government equipment and parts of a small supersonic wind tunnel acquired among the 'reparations' from Germany, he set up his laboratory and in 1949 started research. During the next three years most of the work which he directed was on supersonic flow, but a beginning was also made on studies of turbulence and aerodynamic noise. These were the main interests which he brought with him to Cambridge.

The next professorial vacancy occurred in 1960, when E. B. Moullin reached the retiring age. He was succeeded by C. W. Oatley.

Charles William Oatley was born on 4 February 1904. He was educated at Bedford Modern School, and St John's College, Cambridge. He obtained first class honours in Part I of the Natural Sciences Tripos in 1924 and then read Physics in Part II. For two years after graduation he worked for a firm of radio valve makers, and from 1927 until the outbreak of war in 1939 served on the staff of the Physics Department of King's College, London, under Professor E. V. Appleton. Throughout the war he worked in the Radar Research and Development Establishment of the Ministry of Supply, which was responsible for the development of all radar equipment for the army. He was at first in charge of the basic research group and then became deputy head under (Sir) John Cockroft. When Cockroft moved over to nuclear work Oatley became responsible for all technical work in the Establishment. In 1945 he was elected to a fellowship at Trinity College, Cambridge, and appointed to a university lectureship in the Engineering Department. He was promoted to reader in 1954. He was chairman of the Radio Section of the Institution of Electrical Engineers in 1954–55 and later became a member of the council of that institution.

Professor Moullin, who had been seriously ill for some time, died at his home in Cambridge on 18 September 1963.

Early in 1960 the Faculty Board decided to press for the establishment of the fifth professorship, in the field of mechanical engineering. The General Board agreed to support the proposal and a report by the Council of the Senate making the case for the new chair was approved without opposition on 28 July. A Professorship of Mechanics was established from 1 October, and on 1 April 1961 the election was announced of D. C. Johnson of Trinity Hall, Professor of Mechanical Engineering at the University of Leeds. Professor Johnson had served as demonstrator and lecturer in the department from 1946 and had been secretary of the Faculty Board for more than a year before his transfer to Leeds in 1956. He took up his professorship at Cambridge on 1 April 1962 but resigned a year later in order to return to industry. He was succeeded by E. W. Parkes, M.A., Ph.D., of Caius College, who took up his duties in the Easter Term 1965.

Edward Walter Parkes was born on 19 May 1926. He was educated at King Edward's School, Birmingham and St John's College, Cambridge. He graduated with first class honours in the Mechanical Sciences Tripos in 1945. From Cambridge he went to the Royal Aircraft Establishment and subsequently to firms in the Hawker Siddeley group, working on the design and analysis of aircraft structures.

In 1948 he returned to Cambridge to do research in the field of classical elasticity. He was awarded the Ph.D. degree in 1950 for work on the stresses in flanged beams and in the same year was appointed a university demonstrator. He was elected a Fellow of Gonville and

Caius College in 1953, tutor in 1957 and became a university lecturer in 1954. His research during this period was partly on dynamic plasticity and partly on the structural effects of repeated thermal loading.

In 1959 Dr Parkes went to Stanford University as Visiting Professor and in the following year returned to this country to start the new Department of Engineering at the University of Leicester. This department was intended to provide the undergraduates with a comprehensive view of engineering, emphasizing the inter-relations of the various branches of the subject rather than their diversity and showing the importance of the social as well as of the physical sciences. Its first building, produced in close collaboration between the architects, Messrs Stirling and Gowan, and the staff of the department, has received international acclaim.

It has long been a source of pride to the department that since the end of the last war not a year has passed without one or more members of the teaching staff being elected to chairs in other universities. Between 1948 and 1965 no less than twenty-one readers and lecturers have left for this reason.[1] The presence of so many ex-members of staff in other academic engineering schools is beneficial to all parties but the loss of so many outstanding men had undoubtedly subjected the department to some strain. To relieve this and to make proper provision for the growth in student numbers and research activity it was clearly desirable to provide more senior posts in the department. The Faculty Board, therefore, asked the General Board, in 1964, to establish three more chairs in engineering. Priority was given to the request for a professorship for Mr J. F. Coales, head of the Control Engineering Group, and the need for this post was accepted by the Council of the Senate. A report by that body dated 22 February 1965 asked for the establishment of the new chair from 1 October and this was approved by the Senate on 20 March.

John Flavell Coales was born on 14 September 1907. He was educated at Berkhamsted School and won an open scholarship to Sidney Sussex College in 1926. He obtained honours in Physics in Part II of

[1] They were elected to professorships in the following universities: Belfast, Birmingham, Cardiff, Glasgow, Leeds, Leicester, London, Liverpool, Manchester, Swansea, Warwick, Adelaide, Laval, McGill, Melbourne, Sydney, Cornell.

the Natural Sciences Tripos when he graduated in 1929. He then joined the Admiralty Department of Scientific Research and for some years worked on radio direction finding, ultra short wave radar and communications. During the Second World War he was responsible for the development of naval gunnery radar, and he received the O.B.E. for his services in 1946. In that year he joined the firm of Elliott Brothers (London), Ltd., as research director and built up laboratories at Boreham Wood which covered a wide field of electrical, mechanical and control engineering. His particular interest was in the development of digital and analogue computers. In 1952 he joined the Engineering department at Cambridge as an assistant director of research and was placed in charge of the research and post-graduate work in Control Engineering. Under his direction the group expanded rapidly, and he was promoted to reader in 1958.

Since his return to Cambridge Mr Coales has done much work for automation and control engineering outside as well as inside the university. He has been an active member of the Institution of Electrical Engineers and was chairman of the Measurement Section in 1954–5. He is a member of council and chairman of the newly constituted Control and Automation Division. He was a vice-president of the Congrès Internationale de l'Automatique in Paris in 1956 and president of the Society of Instrument Technology in 1958–9. In 1960 he led the British delegation to the Moscow Congress on Automatic Control and edited the English version of the proceedings. He was a member of the Executive Committee of the National Physical Laboratory from 1958 to 1963, when he spent a year as visiting Mackay Professor of Electrical Engineering in the University of California, Berkeley. He became president of the International Federation of Automatic Control in 1965.

The teaching and research staff

The first new post for which Professor Baker asked after his appointment was that of Superintendent of the Workshops, which had been in abeyance since 1940. The request was approved by the General Board and Mr J. H. Brooks was appointed on 1 February 1946. Under his direction the workshops were entirely reorganized and re-equipped. They have played an increasingly important part in supporting the expansion of research in the department, as well as in producing special equipment for other departments in the university and some outside research institutions. They also provide courses in workshop practice for a limited number of the undergraduates. Throughout his tenure of office Mr Brooks has been supported by Mr A. A. K. Barker,[1] who was promoted to technical officer in 1962. As the capacity of the workshops increased the need was felt for additional help in the design and production of equipment and three new posts were established, that of design engineer in 1957, of electronic design engineer in 1962 and that of a second design engineer in 1965. The total staff of craftsmen in the workshops, from apprentices to higher technical assistants, has been built up steadily, and in the Easter Term 1965 numbered about eighty.

Almost ninety years have passed since the workshops were founded by James Stuart and fourteen less since they became the main cause of his resignation. They are still fulfilling three of the functions for which he designed them, though on a far greater scale than he could ever have imagined: the manufacture of scientific apparatus for the university, the instruction of students in workshop practice and the increase of the departmental income.

The inquiry into the duties and stipends of the university staff which opened in 1943 led to important changes in the organization of university teaching, its relation to college supervision and the system of payment. The first of many General Board reports on the subject was published in January 1947. It proposed that a university teaching officer should be required by ordinance 'to devote himself to the advancement of knowledge in his subject, to give instruction to students, and to promote the interests of the University as a place of education, religion, learning and research'. For this he was to receive a prime stipend with reductions in respect of paid fellowships and college offices. Later the amount of college teaching which he could do without loss of stipend was restricted, and in certain Faculties, including Engineering, a supplementary payment was made when college supervision was further

[1] Nephew of A. W. K. Barker, who was head of the Instrument shop under Professors Hopkinson and Inglis.

restricted in order to give more time for the officer's laboratory work. Throughout the years of this great review, the first stage of which was completed towards the end of 1949, the General Board kept in close touch with the Faculties and as far as possible obtained their prior agreement to changes in the Ordinances. The Departmental Committee on Stipends which maintained this liaison on behalf of the Faculty Board of Engineering consisted originally of Professor Baker, Mr T. C. Wyatt and Mr Landon, who, on his retirement in 1944, was replaced by Dr R. D. Davies.

Professor Baker's first aim was to increase the establishment of lecturers and demonstrators to forty-five in order to improve the student–staff ratio, reduce the teaching load on individuals and allow more time for research. This establishment was achieved in 1950, when the undergraduate population appeared to be stabilizing at about 600, the figure stated by the Faculty Board in 1944 to be the maximum acceptable in the existing buildings.[1] 1950, however, was the year in which the government began its drive for the encouragement of higher technology, and from that date the department entered on a new period of expansion. In 1953, when the scale of assistance likely to be made available to Cambridge was made known by the U.G.C., the General Board invited the departments concerned to make their proposals for staff increases as well as for new buildings and equipment. At its meeting on 15 October the Faculty Board of Engineering asked for 6 readers (including one for Industrial Management to be paid for out of a separate fund), 4 assistant directors of research, 6 assistants in research, 10 university lecturers and 16 university assistants. These proposals were accepted by the General Board and the grant to cover the additional stipends and wages was paid by the Treasury in three instalments in 1955, 1956 and 1957.

The continued rise in the number of undergraduate and research students which will be described in the next section made further increases in the teaching and assistant staffs essential. Applications for new posts were included in every statement of needs submitted annually to the General Board, and by the end of the academic year 1964–65 the approved establishment stood at 6 professors, 6 readers,

[1] See graph on p. 256.

54 university lecturers, 2 university demonstrators, 7 assistant directors of research, 4 senior assistants in research, 5 assistants in research (who do not rank as teaching officers), a technical officer, a secretary, two design engineers and an electronic design engineer, a total graduate staff of 89. At the same time the establishment of university assistants was 180, of whom 104 were in the Scientific and Technical Division, 32 in the Clerical and Library Divisions and 44 in the Maintenance and General Division. In the statement of needs for the quinquennium 1967–72, submitted during the Long Vacation of 1965, further graduate and assistant posts were asked for in order to reduce the student-staff ratio below the present figure of about 11:1.

On 1 November 1965 the Council of the Senate published a report recommending the establishment on a permanent basis of seven new professorships plus two others which had been established for a single tenure only and ten single-tenure professorships. Three of the permanent professorships were to go to Engineering.

The simultaneous establishment of so many senior posts was an event without precedent in the history of the university. It was due to a number of factors to which the Faculty Board of Engineering had drawn attention when, in 1964, they asked for three more chairs in the department. The Council stated in their report that comparison with other universities showed that the ratios of professorial to non-professorial staff to undergraduates and to graduate students were substantially lower in Cambridge than elsewhere. This had resulted in a growing tendency for teaching officers of professorial calibre to take up appointments in other universities and the effect on both teaching and research was becoming increasingly serious. The report was approved without opposition on 11 December.

At the time of writing the precise subjects of the three new professorships in Engineering have not been decided. The Boards of Electors, with common membership, have been appointed and it is hoped that appointments will be made early in 1966. The establishment of these posts will benefit the department in many ways, but will inevitably involve changes in internal organization which cannot yet be foreseen.

The students

In January 1944, when the General Board was preparing its statement of post-war needs, Faculty Boards were asked to estimate the maximum number of students who could be accommodated in existing buildings when conditions returned to normal. The Faculty Board of Engineering replied that they would be able to take 600 undergraduates—45 more than the number in residence in the last year before the war. No reference was made to research students, whose numbers in Inglis's time had varied between 5 and 20, but Professor Baker's plans for the expansion of research activity and the institution of post-graduate courses[1] clearly implied a large increase. By the Michaelmas Term 1954, when the Government's proposals for the expansion of higher technology were accepted by the university there were 664 undergraduates and 45 research students in residence, besides four men who were taking the post-graduate course in structures.

Engineering was not the only scientific department whose post-war development was greater and more rapid than had been expected, and in 1955 the General Board made a special study of the problems created by this expansion. On 28 November 1955 they issued their first report on the development of the university, in which the existing position was reviewed, suggestions for controlling the rate of expansion were examined, and proposals were made for the quinquennium 1957–62. An Appendix to the report contained tables of staff and student numbers in the various faculties. The student–staff ratio in Engineering in 1954 was given as 11·4 compared with 5·8 in the Natural Sciences and Medicine.

In the discussion of the report on 31 January 1956 Professor Baker spoke at some length and strongly contested a suggestion made in the report that 'further expansion in the teaching of applied science and technology might best be left to other Universities, particularly those in industrial areas'. This statement appeared to him to be a gratuitous blow to those who were working in this country with an eye to national needs and to the development of a proper balance in university education. Speakers from other Faculties also found much to criticize and a

[1] See pp. 184 and 241–242.

second report was published on 7 March 1956. In the introduction to the new report the view was expressed that some of the criticisms voiced in the discussion were based on misconceptions and in these cases fuller explanations were given. No further reference was made to the curtailment of technology, and it was stated that: 'As part of the University's contribution to the national need for the training of a considerably increased number of students in science and technology, by far the greater part of the building resources available to the University for major works in the post-war years have been devoted to new buildings for the Departments whose work falls in these fields.' It was further stated that the new buildings already put up for Engineering could accommodate 730 undergraduates and that there would be room for 30 more when the south wing, approved for commencement in 1957, was completed.

In the six years which started in October 1957 the expansion of undergraduate numbers was greater than at any comparable period in the department's history. The annual entry of freshmen, which for three years had been steady at about 236, jumped to 277 in 1957 and to a peak of 286 in 1959. Two other factors led to a further increase in numbers—the decision of the War Office to allow the regular army officers to stay for three years instead of two[1] and the introduction in 1959 of the course in the Principles of Industrial Management,[2] which attracted a number of third year men from other faculties. As a result the total number of undergraduates in the department rose to an all-time record of 839 in 1959. This was also a peak year for the university as a whole, as is shown on the graph on p. 256. With 839 reading engineering out of an undergraduate total of 7,575 the percentage was just over 11 against the normal average of about 10 per cent, which had been maintained with remarkable regularity since the years before the First World War. From 1960 to 1963 the number of undergraduates in the department declined, but in 1964 the upward trend was resumed and in 1965 the intake of 303 freshmen was the highest ever recorded. The build-up of research students during these years suffered no set-back. In the Michaelmas Term 1957 there were 56 research students in residence besides 18 taking the post-graduate courses. Figures for

[1] See p. 231. [2] See p. 235.

these courses remained fairly steady, but the number of research students rose year by year to 79 in 1960, 111 in 1962 and 140 in 1965.

In May 1962, in the Council of the Senate's report on the further development of the Scroope House site[1] the Faculty Board of Engineering was quoted as saying that in the foreseeable future accommodation should be provided for 1,000 undergraduates and about 200 research and post-graduate students. These figures have been accepted by the university and are being used as a target for future planning.

UNDERGRADUATE TEACHING, 1943–1965

Courses and examinations

Undergraduate teaching in the Engineering Department during the 1930s was generally considered to be well adapted to the requirements of both education and industry. The war, however, led to changes which would inevitably affect the qualifications needed by professional engineers, and at his first Faculty Board meeting Professor Baker secured the appointment of a Reorganization Committee to consider the revision of all the courses and examinations in the department. The investigation started immediately, but nearly two years elapsed before any formal reports were submitted to the university. The delay was partly due to the impossibility of effecting any major reorganization during the war, partly to the fact that the end of hostilities brought with it many problems which required short-term rather than long-term solutions, and partly because Professor Baker wished to carry the teaching staff with him before making any major changes in policy.

During the war university courses had been limited to two years, but in the summer of 1945 the Ministry of Labour and National Service extended the normal period of deferment to enable technical students to reside for three years before taking a degree. In 1946 the first releases from the armed forces took place, and the flood of applications for matriculation soon became so strong that the colleges were forced to limit the numbers accepted to read the more popular subjects, which included engineering. Some members of the pre-war staff of the department were still serving, and with the rapid rise in undergraduate numbers due to increased admissions and the reintroduction of three-

[1] See p. 197.

year courses the lecturers in residence found themselves very hard pressed. One by one, however, the absent members returned, new appointments were made, the pressure was somewhat reduced and constructive planning became possible. The views of the staff on the need for changes in the syllabus and examinations gradually crystallized, and in due course the Reorganization Committee was able to bring its recommendations before the Faculty Board for formal consideration.

The first examination to be dealt with was the Mechanical Sciences Qualifying Examination. This had been established by Hopkinson in 1906 in order to separate candidates for the tripos from those who were better suited to read for the Special Examination in Mechanism for the Ordinary Degree.[1] It originally contained papers in mathematics and mechanics only, but a paper on physics was now added and some knowledge of the calculus was made obligatory.[2] Under the new regulations, which were approved in June 1945, candidates were required to pass the examination before the end of their first Lent Term instead of at the end of their first year. This enabled tutors to relegate the weaker men to the Studies Course in time for them to start work on a new syllabus at the beginning of their second year.

As the demand for university education increased in the post-war years colleges raised their entry standards, and the Qualifying Examination tended to become a competitive college entrance examination rather than a separator of Honours from Ordinary Degree men. It was rather different in character from the other examinations in mathematics and mechanics for which the sixth forms in schools were grooming their university candidates and was therefore unpopular with many schoolmasters. No similar barrier existed for candidates for mathematics and the natural sciences, and as time passed it came to be regarded by an increasing number of the departmental teaching staff as an anachronism which had outlived its usefulness. The Examination in Engineering Studies was held for the last time in September 1963,[3] and the original function of the Qualifying Examination disappeared.

[1] See p. 135.

[2] Experience showed the physics paper to be unnecessary, and it was abolished in June 1954.

[3] See p. 221.

Two months later the Faculty Board published a report recommending its abolition, and this was approved on 8 February 1964.

Shortly before the outbreak of war Landon had proposed the substitution of a two-part tripos for the A and B papers, and this principle was accepted by the Reorganization Committee at its first meeting on 8 November 1943, but no detailed plan was submitted while the war was in progress. However, the news that the 1944 entry was to be allowed a third year at Cambridge, which was received during the Long Vacation of 1945, made immediate action necessary. The men concerned were half way through a two-year course for the Tripos Examination, and a quick decision had to be made on the best way to employ the third year. It happened that the 1944 entry contained an unusually high proportion of men of first-class ability, and two members of the committee agreed to work out a course leading to an examination which would constitute a Second Part to the tripos based on selected B papers. The course was an outstanding success, and in the first examination in 1947 no less than 24 of the 81 candidates obtained first class honours.

The Reorganization Committee's proposals for a two-part tripos had been accepted at a special meeting held on 8 October 1945. Part I was to cover much the same ground as the existing A papers, while Part II would consist of more advanced questions arranged in four groups representing civil, mechanical, aeronautical and electrical engineering, and would replace the B papers. The best men would be given a course (to be known as the 'fast course') leading to Part I of the tripos at the end of their second year, and this would be followed by a one-year course of a more advanced and more specialized nature leading to an examination in one of the groups of subjects in Part II. Most of the men, however, would do a 'normal' course and take Part I at the end of their third year. They could then, if they wished and if their college agreed, take Part II at the end of a fourth. Honours in Part I would qualify for the B.A. degree, but a man who took it after two years would require a third year's residence in which he could either take Part II or work for a Certificate of Diligent study. Entry to the fast course would be restricted to men who were placed in the first or second class in the Second Year Preliminary Examination.

The General Board objected to the restriction on entry to the Fast Course, but this was a matter of principle to which Professor Baker and the Faculty Board attached great importance. They were determined that the B.A. degree with honours should continue to imply a broad education in fundamental principles and that nothing should be done to create an impression that success in Part II was a normal requirement for a Cambridge engineering graduate. Furthermore, the division of tripos candidates into two courses would make more efficient teaching possible and would be greatly in the interests of the men themselves. It was finally decided that no restriction on entry to the fast course would be embodied in Ordinances, but that tutors would be urged to admit only their ablest men to it. The report, amended in this sense, was published on 13 November and approved a month later.

Four years were allowed to pass before a Schedule for Part II was embodied in Ordinances. During this period experience showed that the form of the examination played too large a part in determining the undergraduate's programme of work. Too many papers were required of him to allow concentration in depth in a limited field, and he was given an unnecessarily wide choice of subjects. In a report published by the Faculty Board on 24 April 1950 the regulations were amended so that a candidate had only to offer four papers instead of six and the number of papers in each group was reduced. An appendix contained the schedule for the Examination.[1]

A second report on Part II published in July proposed that the examination should not be classed, but that the names of successful candidates should be published in alphabetical order as is the practice in Part III of the Mathematical Tripos. A majority of the General Board opposed the proposal on the grounds that it was contrary to normal practice and that the educational issues were debatable, and as negotiation failed to produce a compromise the unusual step was taken of printing the Faculty Boards' report with a note attached to it signifying General Board opposition. As the examination is still unclassed the reasons which led the Faculty Board to maintain its stand are worth

[1] Papers in Part II were as follows—Group A: Strength of Materials I* and II,* Theory of Structures (2 papers), Civil Engineering, Hydraulics. Group B: Strength of Materials I, Mechanics of Prime Movers,* Thermodynamics I* and II (2 papers), Mechanisms and the Mechanics of Machines. Group C: Aerodynamics (3 papers), Aeronautical Engineering. Group D: Electricity (4 papers). [* Denotes compulsory paper.]

considering. The arguments were developed at length in the report and further elaborated in the discussion on 16 January 1951. They may be summarized as follows.

Part II is concerned with the application of science to some of the more complex problems of engineering, and aims at giving the candidate a thorough understanding of the fundamental principles and methods of one chosen branch of the subject. The need for an examination at the end of the course was accepted by the Faculty Board, but it was argued that if candidates were not classed they would feel freer to work in a wider and deeper field than the usual one without jeopardizing their academic record. Experience showed that it was difficult to set an examination which divided Part II candidates fairly on a basis of their intellectual ability. The papers had to strike a balance between mathematical questions and those which involve discussion of physical principles and engineering methods, but engineering provides few realistic problems which can be solved completely in an hour or so. It is possible to devise artificial problems for examination purposes, but this type of question may tempt the student to rely on the clever or experienced coach or supervisor to the detriment of regular teaching and his real education. It was also felt that an unclassed Part II reduces the risk that men who have only taken Part I will be thought by industry to have done only half the course.

The report was non-placeted when it came up for approval on 9 February, but was approved by a sufficient majority to make voting unnecessary.

The Faculty Board's first report on the Studies Examination was published on 26 November 1945. The introductory paragraphs explain the need for a new type of course.

There is need for an engineering course other than the Tripos course, because there are many students who, while they lack the mathematical ability for the Tripos, will yet with suitable training make useful Engineers. But, owing principally to two defects, which were apparent before the war and are likely to be more so in the changed conditions after it, the existing course and examinations in Engineering Studies do not satisfactorily fill this need. The first defect is that, while difficult theory is omitted its place is not adequately filled by more practical instruction within the capacity of the students; as a result the course is incomplete and in addition the students have too little to do. The second defect is that a student can only take such subjects as Electricity, Heat, or Structures for one of the examinations so that he only studies these subjects for eight months, although his Certificate of Proficiency is endorsed as if he were a specialist in them.

One indication of the unsatisfactory nature of the course is that Directors of Studies are on the whole so loath to relegate students to it that the Tripos course is burdened with weak men. Another indication is that two of the principal Engineering Institutions no longer accept the Certificate of Proficiency as conferring exemption from their entrance examinations.

The Board consider that a change of system is needed and that the course should be re-organized as a continuous three-year course with a fuller programme and a higher standard. Specialization should only be allowed in the third year, but much of the groundwork of all branches should be covered in the first and second years. Such a course should be designed to provide suitably for the weaker students of the present Tripos course and for the majority of the students who take the present course in Engineering Studies. The Board consider that the weakest students in the present Engineering Studies course have not the ability to profit from a university course in engineering, and that no attempt should be made to provide for them.

The new examination was to be in two Sections, the first to be taken not earlier than the fourth term of residence and the second not earlier than the seventh. As this was not a tripos examination candidates were allowed more than one attempt at it, and examinations in both Sections were to be held twice a year. Success in both Sections would lead to a Certificate of Proficiency. The report was approved without opposition on 9 February 1946 and the courses were in full operation by October 1948. A choice of papers in the examination for Section II allowed some specialization during the third year.[1] A Progress Examination in Engineering Studies, to be taken at the end of the first year as an equivalent to the Preliminary Examination for the tripos was established in May 1947.

The Studies Course was an attempt to create something quite new

[1] Section I of the Studies Examination had papers in Mathematics, Applied Mechanics, Heat, Structures, Electricity, Hydraulics, Workshop Practice and Materials, and Drawing. Candidates for Section II had to take papers on Industrial Administration and Economics and Drawing (Schedule IIA) and three out of the following papers (Schedule IIB): Theory and Design of Machines, Metallurgy, Civil Structures or Electricity (1), Surveying and Geology or Electricity (2), Heat Engines or Electricity (3) and Hydraulics or Aeronautics. Candidates were also required to prove their competence in practical surveying and laboratory and drawing office work and furnish evidence of workshop experience.

in the department. It was neither a re-furbishing of the old Engineering 'Specials' nor a watered-down edition of the tripos, but was planned with a limited but clearly defined object, and it was hoped that in time the examination would win for itself a reputation as highly esteemed as that of the tripos, though totally different in kind. The course had a warm welcome from the engineering profession, and for a year or two it seemed possible that it would achieve its object by attracting a sufficient number of able candidates whose minds were not suited to the particular discipline of the tripos. Gradually, however, as pressure for entry into the university increased, the colleges became less willing to admit men to read for any course which did not lead to a degree with honours, and instead of attracting the type of men for whom it had been intended it became a refuge for those who had been put down from the tripos course through lack of sufficient mathematical ability or because they did not work hard enough. In 1960 the Faculty Board regretfully recommended that the course should be abandoned, and their report to this effect was approved on 24 June. Section I of the examination was held for the last time in September 1962 and Section II a year later. It had been sacrificed on the altar of expediency, to the great detriment of technological education to which, if all had gone well, it might have contributed something unique and of real value.

Almost a century had passed since the introduction in 1865 of the Special Examination in Mechanism and Applied Science had first made engineering a subject for the Ordinary Degree.

The separation of the tripos candidates into fast and normal courses worked well, but year by year new material was included in the lectures and laboratory work and by 1955 it was becoming evident that all but the most able candidates were being asked to learn too much. The Faculty Board decided to review the courses and examinations for which it was responsible with the primary object of cutting out from the syllabus everything which was obsolete or non-essential. They were greatly aided in this task by the subdivision of the teaching staff into groups. With the founding of new chairs and the establishment of more teaching and research posts some measure of decentralization had become essential, and groups under the chairmanship of the appropriate professor, reader or lecturer were by this time well established.[1] Meetings were held periodically to discuss developments in teaching, new fields of research, bids for new equipment, or other matters of common interest, and the groups were therefore well qualified to examine critically the teaching and examination syllabus in their own subject. Discussions within the groups were followed by meetings of heads of groups and by the end of the Easter Term 1956 it became possible to draw up a comprehensive list of recommendations for consideration by the Faculty Board. These included the introduction of a Combined Preliminary Examination for the first-year fast course, changes in the marks awarded for various papers and a significant reduction in the total number of lectures. The General Paper was given more importance than formerly and divided into two parts, the first containing questions on lectures given in the department and the second being an essay chosen from a short list of subjects. To help students to increase their general knowledge and acquire new ideas the Faculty Board was to approve a reading list which would be revised annually. Laboratory teaching was examined with the same thoroughness as the lecture programme and several changes were proposed. All the recommendations listed were approved by the Faculty Board on 21 May and implemented during the course of the next academic year.

Other changes were made in November 1958, when the regulations for the tripos and the Certificate of Proficiency in Engineering Studies were again revised. The second class in the honours list of the tripos examination was divided into two parts, and practical courses on Electronic Instrumentation and Metrology were introduced as alternatives to Surveying. All candidates for the tripos are required to complete one of the three courses.

The review of the Part I syllabus which began in 1955 brought into the open substantial differences of opinion among the staff about the teaching of electrical engineering. All academic courses for a first degree have to strike a mean between a syllabus which is too wide to constitute

[1] See p. 200. In 1965 the groups were: Structures (including Soil Mechanics and Stress Analysis), Aeronautics and Fluid Mechanics (including Hydraulics), Electricity, Thermodynamics, Mechanics, Mathematics, Drawing, Control Engineering (post-graduate only) and Industrial Management. In certain cases teaching officers were members of more than one group.

an effective intellectual discipline and one which is too narrow and specialized to provide a proper education. The problem is particularly difficult in the case of engineering, where the field covered by the profession is so vast and the knowledge required by engineers so diverse and yet so exact. United Kingdom universities are almost alone in limiting the engineering degree courses to three years, though not all of them do so. Most of the old-established provincial universities have separate schools or departments for the various branches—civil, mechanical, electrical, etc. Cambridge, for reasons which it is hoped have been made plain to the reader, has always attempted to defer specialization until a broad education has been given in the general principles which underlie all branches of engineering, and this example has been followed by some of the new universities.

The wish to maintain a single syllabus for Part I was deep-seated in the department, but some members of the Electrical group, notably the electronics specialists headed by Mr C. W. Oatley, contended that the existing course no longer provided an adequate basic training for future electrical engineers. Research in physics and technology had proceeded at an unprecedented pace since the war, and principles which previously had little application to engineering were now of the greatest practical importance. The Faculty Board always tried to keep the undergraduate courses in line with developments in engineering science, but pressure on the syllabus was intense, and when making changes they were forced to be selective. In consequence certain aspects of modern physics with applications in electrical engineering had failed to win a place in the Part I syllabus. One result of this was that candidates offering electricity in Part II were faced with an overcrowded curriculum, and much of their time was necessarily devoted to elementary parts of the subject which would have been more appropriate in Part I. In 1958 Mr Oatley suggested that options should be allowed in the second-year fast and third-year normal courses and in the tripos itself, so that those who intended to become electrical engineers could receive instruction in the developments in physics and allied subjects which were essential to the study of electronics, but not for other branches of engineering, The dilemma was evident to everyone, but the desire to maintain an undivided tripos was predominant, and when a vote was taken there was a decisive majority against the proposals, and for the time being the matter was dropped.

However, in October 1960 Mr Oatley became Professor of Electrical Engineering and he at once returned to the attack. Within a month of his appointment he had put on paper detailed proposals for the institution of a separate Electrical Sciences Tripos which the Electrical Group unanimously agreed should be circulated for further discussion, although about a third of them did not favour the proposals as they stood. The report listed the inadequacies of the existing Part I course for prospective electrical engineers and suggested a schedule for a new examination, the Electrical Sciences Tripos, Part I, to be taken at the end of the second fast or third normal year, and a Part II to replace the electrical option in the Mechanical Sciences Tripos, Part II. The report was discussed at a general staff meeting on 5 December, and as opinion was sharply divided a Teaching Committee, recently set up by the Faculty Board, was asked to consider the problem and make recommendations. On 3 February 1961, the committee circulated two alternative schemes—Professor Oatley's proposals for a two-part Electrical Sciences Tripos and a compromise suggestion by the secretary of the Faculty Board. Both were discussed by the Faculty Board on 27 February and considered by another full staff meeting on 13 March, but no solution to the problem could be found. Discussions and negotiations continued throughout the summer, and agreement was finally reached on a report which was signed on 23 October with only one member of the Board abstaining. The report was published on 11 January 1962 and approved on 10 March.

The attempt to establish an Electrical Sciences Tripos in two parts was abandoned, but men who are reading for the Mechanical Sciences Tripos, Part I, can now choose between two options—the General and Electrical—which are offered to normal course students in their third year and to fast-course students in the Lent and Easter Terms of their second year. The General option is similar to the old course for Part I while the Electrical option reduces the non-electrical content somewhat in order to make room for the new subjects now needed by electrical engineers. The same options are allowed in the examinations. It must be stressed that all the subjects in the syllabus are covered in both options. The electrical option in Part II has been renamed the Electrical Sciences Tripos, but its candidates, unlike those in the Mechanical Sciences Tripos Part II, are classed.

Examinations under the new regulations were held for the first time in 1963. In the examination for Part I 24 fast course men out of 96 and 20 normal course out of 172 took the Electrical option. Comparable figures for 1964 were 25 men out of 103 and 17 out of 156 and for 1965 29 out of 96 and 28 out of 148.

A recent development which may ultimately have considerable repercussions on the undergraduate teaching syllabus is the growing realization of the importance in engineering of the role of the computer. For some years courses have been given in computer programming, using equipment in the Mathematical Laboratory. Funds have now been obtained for the purchase of a departmental computer suitable for all the undergraduate teaching in this subject and it is hoped that it will come into use in the academic year starting in 1966.

Courses for serving officers

The university's first course in peacetime for regular serving officers started very soon after the First World War ended. Within a matter of weeks after the signing of the Armistice the Admiralty obtained the agreement of the university to the institution of a six months course for junior officers who had been sent to sea before completing the education they would normally have received at Dartmouth. The course consisted of lectures and supervision in mathematics, electricity, applied mechanics, naval engineering and similar subjects with the option of attending a course of more general interest. Most of the colleges agreed to accept a quota of the young officers, and after the first term all were matriculated. Captain E. J. A. (later Admiral Sir Eric) Fullerton was appointed in command of the course, which started work in the Lent Term 1919. Much of the teaching was given in the engineering laboratories and lecture-rooms by the departmental staff. Mr Donald Portway, a lecturer in the department who afterwards became Master of St Catharine's, kept a friendly, though unofficial, eye on them. He had been on the staff of the Naval College at Dartmouth for three years before the war and for two terms after the Armistice, though he was in the Royal Engineers throughout the war and saw much active service with them in France. Captain H. W. (later Admiral Sir Herbert) Richmond, who became Master of Downing College after his retirement, was Director of Training on the Naval Staff when the scheme was inaugurated. Four hundred sub-lieutenants were in residence during the Lent and Easter Terms of 1919, but the number in each course was later reduced to about 140.

It was inevitable that the presence of so many high-spirited young officers fresh from the dangers and hardships of life at sea in wartime should cause difficulties, but with good will on both sides these were largely overcome. Rudyard Kipling wrote a poem about the first batch of naval officers in Cambridge which he called *The Scholars, 1919*. Part of it ran as follows:

Far have they steamed and much have they known, and most would they fain forget;
But now they are come to their joyous own with all the world in their debt.

Soft, blow soft on them, little East Wind! Be smooth for them, mighty stream!
Though the cams they use are not of your kind, and they bump, for choice, by steam.
Lightly dance with them, Newnham maid—but none too lightly believe.
They are hot from the fifty-month blockade, and they carry their hearts on their sleeve.
Tenderly, Proctor, let them down, if they do not walk as they should:
For, by God, if they owe you half a crown, you owe 'em your four years food.[1]

This kindly and tolerant attitude was undoubtedly shared by the university and college authorities. Judging by the stories which are still recounted with relish by the few surviving college servants who remember those days there were occasions when quite a lot of tolerance was needed.

The scheme was revised during the Long Vacation of 1919 in the light of experience gained during the first course and then remained substantially unchanged until it was terminated in the summer of 1923. The 'Specified Course' consisted of lectures on Mathematics, Physics (Electricity, Optics and Sound), Engineering (Applied Mechanics and Applied Electricity), Marine Engineering, Navigation and a War Course comprising one lecture a week. Officers were also allowed to choose

[1] *Rudyard Kipling's Verse* (Definitive Edition, Hodder and Stoughton, London, 1940).

subjects from an 'Extra Course' in which lectures were offered in English Literature, Naval History, the Social and Political History of the British Empire, French and other modern languages, the History of Geographical Discovery, the Natural History of the Sea, Ethnology (savages past and present) and the Technical Application of Chemistry. Only the lectures on Engineering, with their associated laboratory and class work, were held in the Engineering Laboratory. While in residence the officers remained under naval discipline but they were also required to conform to all rules and regulations issued by the Proctors and college authorities. 'Unit Commanders' of commander's rank were appointed to each group of two or three colleges and in collaboration with the college authorities they dealt with all offences which seemed to merit punishment. One private of marines was supplied to each ten 'or fraction of ten' officers to valet and look after their clothes. One hundred and nine officers came up for the second course in October 1919. Among them was the present Admiral of the Fleet Earl Mountbatten of Burma, who was attached to Christ's College.

The Cambridge course was naturally very popular with the sub-lieutenants, but it was no less highly valued by the Board of Admiralty, and between 1919 and 1923 great efforts were made to continue it in some form for selected officers of somewhat higher rank. Various proposals were made and examined, but in the end they all had to be abandoned because the Treasury was unable to allocate the money required to finance them. The 'Geddes Axe' fell on the Navy in 1922 and the Cambridge scheme was one of its victims. Since the first batch came up in January 1919 over twelve hundred naval officers had attended the courses.

The First Lord of the Admiralty visited Cambridge in February 1923 and had the 'unique experience of inspecting ranks of naval officers arrayed in academic dress'. When the last officers had gone down the Second Sea Lord wrote a letter of thanks to the Vice-Chancellor in which he said:

> Apart from the study and the educational advancement as tested by examination, it is realised—and by none better than by the Officers themselves—that they have derived a lasting benefit in other ways from their brief contact with University life. After the strain, monotony, and the disturbing effect of war service generally,

their sense of proportion has been restored, and a new outlook on life has opened. They have carried away with them not only the happiest recollections of their life at Cambridge, but something of that indefinable and permanent influence which the University has on all her sons.[1]

The Royal Air Force was the first of the three services to send regular officers to take the full course for the Mechanical Sciences Tripos. In the Michaelmas Term 1919 ten technical officers were admitted by the colleges, and as far as circumstances permitted no distinction was made between them and the civilian undergraduates. No R.A.F. freshmen came up in 1920, but from 1919 to 1937 the department was never without its regular R.A.F. officers. The number in residence varied between nineteen and two. Among them was Flight Lieutenant, now Sir Frank Whittle, who obtained first class honours in the Mechanical Sciences Tripos in 1936 and then stayed on for a year's post-graduate study.

In the Michaelmas Term 1920 fifty officers of the Royal Engineers came up for a course, and this was repeated with similar numbers in the next three years. In the three following years numbers were approximately halved. In 1927 a special two year course for the tripos, restricted to army officers, was started, and this continued until the outbreak of war in September 1939. Most of the first-year lectures were omitted from the course and the R.E. freshmen were classed in the department as 'second-year students'. The officers took the examination for Part I of the Mechanical Sciences Tripos at the end of their second year of residence on the same footing as the third-year civilian undergraduates. A high proportion of them obtained first class honours and the reputation of the Royal Engineer officers stood high in the university and the colleges. Between 1927 and 1934 their numbers varied between twenty-one and thirty-one in residence but in the four years before the outbreak of the Second World War they rose to between 35 and 46. From 1929 onwards two officers of the Royal Corps of Signals were also sent up each year.

During the Second World War no regular serving officers took the course for the tripos, but a number of courses lasting six months were

[1] The author is indebted to Captain S. W. Roskill, D.S.C., M.A., R.N. for permission to make use of his article 'The Navy at Cambridge, 1919–23', published in *The Mariner's Mirror*, vol. 49, no. 3 (August 1963).

arranged for cadets of the three services. Sixty-five R.E. cadets came up in the Michaelmas Term 1940 and 58 R.A.F. cadets in the following Easter Term. The Royal Engineers were joined in the Easter Term 1943 by cadets of the Royal Marines and the Royal Army Service Corps, bringing the total number in the course to 123. Royal Air Force numbers had risen to 100 in the Michaelmas Term 1942 and they remained high until the fighting in Europe ended two years later. Courses for the Royal Engineers and the Royal Air Force continued until the war against Japan ended in the late summer of 1945. Three six-month courses for naval cadets were given. Thirty-six cadets came up in the Easter Term 1943, 40 in the Michaelmas Term of the same year and 36 in the following Easter Term. All the service courses which started in the Easter Term carried through into the Long Vacation.

When the war ended the Army and the Royal Air Force asked that their pre-war practice of sending selected officers to read for the tripos should be resumed. At about the same time the Royal Navy began to enter cadets into the new Electrical Branch, and the three year course for the Mechanical Sciences Tripos was included as part of their training. It was hoped that most, if not all, of these officers would qualify for the fast course which had just been instituted and would thus be able to specialize in electrical engineering during their third year. Experience showed, however, that relatively few were able to cope with the fast course and many had to be content with the electrical option in the course for Section II in the examination for the ordinary degree.

These requests for regular annual entries of officers from the three services would, if approved, have put a heavy strain on the capacity of the department and the colleges at a time when an unusually large number of civilians were applying for admission to read engineering, and in 1947 an Inter-Service Training Committee was set up to study the problem. After lengthy negotiations it was agreed that the department would allocate not more than 120 places to serving officers out of an expected total of about 600 undergraduates—one officer to every four civilians. The freedom of the colleges in the matter of admissions was formally recognized, but it was accepted in practice that the Royal Navy and Royal Air Force would be allowed to enter 20 and 6 men respectively each year for a three-year course and the Army 20 for a two-year course. As a result the Service population in the Michaelmas Term 1950 amounted to 118 out of a total of 591 undergraduates in the department.

Naval entries during this period were confined entirely to officers of the Electrical Branch. Most of the Army officers belonged, as formerly, to the Royal Engineers and Royal Corps of Signals. The Royal Air Force ran a special scheme of direct entry for a small number of ex-apprentices who had undergone preliminary training at the R.A.F. Technical College at Henlow. In 1952 they introduced an expanded university cadetship scheme of the sandwich type under which cadets were to serve for a year at Henlow before matriculating and go back there for a further eight months after taking their degree. They asked that their annual allocation to Cambridge should be increased from six to twenty, and the Inter-Services Training Committee supported their request after ascertaining that the Engineering Department could take fourteen additional students, but the Council of the Senate felt that the time had come for a reappraisal of the whole question of service entries, and referred the matter to a Special Committee. The committee expressed some doubt, which was shared by an increasing number of people in the university, whether it was in the national interest that so large a proportion of places in one of the foremost engineering schools in the country should be allocated to members of the armed forces. They pointed out that the quota system instituted after the war was no longer in force, and that the colleges could not be committed to accept a fixed number of serving officers each year. The Tutorial Representatives of the Colleges[1] and the Engineering department were consulted and it was eventually decided by the Council that serving officers would in future have to compete for admission on equal terms with civilians wishing to read for the Mechanical Sciences Tripos. This ruling was accepted by the Service Ministries, but for some years admissions continued at much the same rate as before.

Over the years, however, the Services made a number of changes in their schemes of technical education. External degree courses in Engi-

[1] An unofficial body which deals with certain matters of common interest to the colleges.

neering were introduced into the naval colleges at Greenwich and Manadon in 1960 and in consequence the Cambridge entry was greatly reduced. Officers destined for the Mechanical as well as the Electrical Engineering sub-specialization were included and as the entry was more selective the average standard of ability was correspondingly improved.

In 1955 the special two-year course for the Army officers was discontinued. Since that time most of the Army officers have taken the three year normal course. Those taking the fast course are usually allowed a third year for Part II if they obtained first or second class honours.

Meanwhile, in the Royal Air Force dissatisfaction was felt with the cadetship scheme, as those who were left behind at Henlow felt themselves at a disadvantage in comparison with those who went on to a university. Furthermore, the scheme only catered for a graduate entry to the Technical Branch and made no provision for the General Duties Officers. In consequence a new scheme was instituted in 1963. Twenty-five cadetships (increased a year later to forty) are awarded each year, half to the Technical and half to the General Duties Branch, either to schoolboys or to undergraduates but not to officers already serving. They can be held at any British university, and applicants must obtain their own places in open competition with all other candidates. Those who wish to join the Technical Branch must read engineering or some other acceptable scientific subject.

Thus by 1964 all the three services had abandoned the privileged position which had been granted to them by the colleges when the quota system was in force after the Second World War. Under their new schemes numbers coming to Cambridge have been greatly reduced but the average quality of the intake has been correspondingly improved. In October 1965 the entries were 5 for the Navy, 15 for the Army and 2 for the Royal Air Force. Numbers of regular officers in residence were respectively 14, 45 and 6.

Courses on Industrial Management

Industrial Management has of recent years been fast gaining recognition as a subject for study in university schools of engineering. Successful engineers have to understand how to handle work groups and organizations as well as materials and machines. Scientific principles of inquiry and experiment which have proved successful in solving problems in the natural and applied sciences can be adopted in the study of human affairs. Engineers are therefore interested in the work of those who have treated social phenomena in industry as systematically as they themselves have treated the behaviour of mechanisms. In the world of affairs the two problems can rarely be separated, and an industrialist who has studied their interactions dispassionately while still at a university may well have an advantage over one who has not done so.

Understanding of the large work organizations characteristic of our society is not a prerequisite for success in all professional careers, but many graduates in engineering eventually find themselves in the manager's chair or at the table of the Board of Directors. People in such positions will benefit if they are able to apply objective methods to the solution of their problems and if they have acquired some familiarity with the basic concepts of personality, group behaviour and the economics of the firm.

Up to the present time much of the academic work in these fields has been done in a number of American universities such as Harvard, the Massachusetts Institute of Technology and the Carnegie Institute. Their importance has, however, long been recognized in Cambridge, and probably to a greater extent in the Engineering Department than elsewhere. Certainly the professors from Stuart onwards have maintained the closest possible links with industry, and have tried to match the education of their undergraduates with the requirements of the times.

Inglis had only been four years in the chair when, in 1922, he arranged for R. W. Stanners of Caius College, a graduate in Economics, to give a course of lectures to third-year students for the tripos on 'Economics of the Metal and Engineering Trades'. The course was repeated every year in the Lent Term until 1936. In 1930 T. G. Rose, an engineering consultant, gave the first course on 'Management' to the same audience. It was so successful that the Faculty Board passed a vote of thanks to him, and the course was continued annually until 1947. At first it was given only to tripos men, but from 1935 a similar course was given to men working for the Third Examination in Engineering Studies. In

1938 Stanners's lectures were replaced by a course on 'Aesthetics of Engineering Design' by Dr Oscar Faber, another consulting engineer, and a friend of Inglis.

Three years later, when Inglis was President of the Institution of Civil Engineers, the Council of that body offered the university an annual grant of £1,000 for five years 'to foster among engineers the closer study of the economics of engineering projects, the organization and management of engineering works and the relations of aesthetic considerations to engineering design and construction'. The stipulation was made that the Engineering department should be responsible for teaching these subjects and for 'such research and study as may be necessary to ensure progress and alignment with changing conditions'. The offer was accepted and a course of evening lectures instituted, open to the public and free of charge. The first lecture was given by Inglis himself, and attracted a large audience.

In 1947 the series was continued on slightly different lines when Professor Baker decided to invite some leading members of industrial firms to give short courses on management and allied subjects. The first speaker was Mr H. G. Nelson, at that time a director of the English Electric Company, now Lord Nelson of Stafford, who had graduated in the Mechanical Sciences Tripos in 1937. He gave five lectures in the Michaelmas Term and repeated the course in the following year. In 1950 seven lectures on 'The Objectives and Principles of Industrial Management' were given by Sir Ewart Smith, Dr R. Beeching, now Lord Beeching, and Mr R. M. Currie of Imperial Chemical Industries, Ltd. Somewhat similar courses have been given by representatives of other industrial companies in succeeding years.

When the new course in Engineering Studies started in 1948, management was included in the syllabus for Section II. Mr A. L. Bird took over the lectures for the old Studies Examination from Mr Rose from 1947 until 1949, when it was held for the last time, but the course for the new Section II was entrusted to Mr D. L. Marples, a graduate of the department who had just joined the staff as a demonstrator. He continued to lecture to the Studies men until the course came to an end in 1963. In the Lent Term 1951 a course on Industrial Relations was given to tripos candidates by Professor H. S. Kirkaldy, Professor of Industrial Relations, and a course on Industrial Psychology by Mr E. H. Farmer of Trinity, a university lecturer on Psychology.

The teaching of Industrial Management was thus well established in the department when, in September 1953, the University Grants Committee, of which Professor Baker was a member from 1954 to 1964, proposed to the Vice-Chancellor that a chair or readership in Engineering should be endowed with funds derived from United States economic aid, and that the holder of the post should develop the fields of study lying between engineering, economics and industrial psychology. The General Board referred the suggestion to Professor Baker, who gave it his full support, pointing out that his Faculty Board had already recommended a readership in the same subject. The other professors concerned also supported the proposal, and on 1 March 1954 the Council submitted a report which recommended acceptance of the endowment provided that the terms of the trust, while restricting use of the income to work in the prescribed field, did not bind the university to the permanent maintenance of any particular post. In the first instance the Council proposed the establishment of a Readership in Industrial Management in the department of Engineering. The report was approved on 20 March and the readership established from 1 January 1955. The appointing body consisted of the Vice-Chancellor, the Head of the Engineering Department, the Professors of Industrial Relations and Experimental Psychology and five other members, two of whom were to be non-residents.

Attempts to find a candidate of sufficient status to qualify for a readership proved unsuccessful so the post was changed temporarily to that of Assistant Director of Research. On 12 December the appointment was announced of Mr F. J. Willett of Fitzwilliam House, with tenure for five years from 1 January 1957.

Mr Willett had graduated in Anthropology in 1948 and after serving in the Fleet Air Arm had had several years managerial experience in industry. Before starting his work in the department he paid a round of visits to some of the American universities which had established a reputation in this field of study and consulted people in industry and other Faculties in the university who were interested in the project. In the Michaelmas Term 1958 Professor Baker, while on a visit to the

United States, spent six weeks at Harvard attending classes in the Graduate School of Business Administration.

While determining what form the teaching of industrial management should take under the new endowment the Faculty Board considered the possibility of establishing a post-graduate course for relatively senior men with industrial experience, but finally decided to confine it in the first instance to the undergraduate or immediately post-graduate level. It was not considered appropriate to include it in the syllabus for Part I, as this is an engineering course and the projected course was inter-disciplinary between engineering, sociology and economics. Advantage was therefore taken of the Part II system to add a new option in the Principles of Industrial Management. Because the course would demand mathematical and statistical skills which Arts men would be unlikely to possess, as well as a high standard of more general intellectual ability, admission was restricted to men who had obtained first or second class honours in the Mechanical Sciences, Mathematical or Natural Sciences Tripos. Later it was opened to graduates of other Universities in the same subjects who possessed equivalent qualifications. To advise the Faculty Board on the conduct and development of the course a committee was established under the chairmanship of the Head of the Department, with three members appointed by the Board and four others from the Faculty Boards of Economics and Politics, History, Law and Biology B. A report on the course and a schedule for the examination were published on 26 January 1959 and approved on 7 March. The first course for the examination started in the Michaelmas Term 1959 and was attended by twenty-six men.[1] Mr Marples and Mr Willett gave most of the teaching, but they received great assistance from lecturers from other Faculties and from professional specialists in industry. The course was repeated on similar lines in the two following years, and in 1961 extension of the research which ran parallel with the teaching was made possible by the establishment of two more assistant directorships of research. The stipends and expenses associated with these posts were covered by funds provided by the Foundation for Management Education supplemented by a grant from the Treasury.

[1] Of these twenty-one had taken Part I of the Mechanical Sciences Tripos, four the Natural Sciences Tripos and one Part I of the Mathematical Tripos.

In the summer of 1962 Mr Willett resigned his post on appointment to the Sidney Mayer Professorship of Business Administration at Melbourne, Australia. Mr Marples carried on the course during the following year with the help of three resident visitors, one from the Social Sciences Research Centre at Edinburgh University, one from the University College of South Wales and one from a firm of consultants on industrial market research.

From 1954 onwards Mr Marples carried out a considerable amount of research and extra-mural work on management in addition to his teaching duties. In the Long Vacation of 1954 and the two following years he, with Mr J. Reddaway, directed experimental courses on post-graduate industrial training in conjunction with the Napier and Glacier Metal Companies.[1] The funds for these courses were provided by the Nuffield Trust, and their aim was to study the possibility of giving a formal training in machine shop technology and management in an industrial environment. Similar courses were subsequently arranged with other firms, and the idea has been incorporated and developed in the graduate training schemes of a number of companies. Mr Marples was also responsible for the design and direction of the first three residential courses for managers in industry which the Board of Extra-Mural Studies had held every summer since 1953 at Madingley Hall near Cambridge. Another experimental course which he designed and directed was held three times between 1961 and 1964 in Churchill College. It was a sandwich course with four residential sessions planned to meet requirements formulated by a joint committee of the Institution of Electrical Engineers and the British Institute of Management.

In September 1963 Dr C. Sofer of the Tavistock Institute was appointed as the first Reader in Industrial Management. He is a sociologist who was trained at the University of Cape Town, the London School of Economics and Harvard. He worked first in the field of community and family studies and later on the comparative study of various types of organization—hospitals, educational and research units and industrial firms. His main recent work had been in the field of conflict resolution and the management of major organizational changes. His appointment was consistent with the developing policy

[1] *Proc. Inst. Mech. Engrs*, vol. 170, no. 22 (1956).

of the university and the department in the teaching of management to treat this as a field of study rather than as a discipline in its own right or as a set of skills. It followed from this that those teaching in the course should be specialists in one or other of the disciplines that are being applied to the study of managerial behaviour.

The syllabus of the course was revised in 1964 and the number of papers in the examination reduced to three in order to deepen understanding and carry to its logical conclusion the policy of centring on disciplines—Social Psychology, Economics, and Mathematical and Statistical Analysis. In addition students are required to write a substantial paper bearing on the relation between industry and society. Teaching now concentrates on the functioning of the industrial enterprise and on its immediate contacts with the environment. The research programme in progress at the end of 1965 was concentrated on empirical studies of industrial organizations. The main topics under investigation were conditions affecting collaboration and conflict between scientists and administrators within large organizations, career preoccupations of executives and professionals and social process in decision making.

RESEARCH AND POST-GRADUATE TEACHING, 1943–1965

It would be inappropriate in a book of this character to give a detailed account of the research which has been carried out in the department since Professor's Baker appointment, but the lines on which it has developed will be briefly described and some of its main achievements high-lighted. Research has always been carried out in the department, but under Inglis undergraduate teaching had been given overriding priority. The acceptance of Professor Baker's proposals[1] by the Faculty Board on 25 September 1943 led to a great increase in research activities without any detriment to the importance given to teaching, but this was only made possible by the expansion of the professorial and teaching staff and the provision of new laboratories. The first developments naturally took place in those branches of civil engineering with which Baker had been most closely concerned before his election to the Cambridge chair.

[1] See p. 183.

Structures and materials

Very shortly after Professor Baker's appointment the Welding Research Council, which had financed some of his later work at Bristol, promised to renew its support and to pay the stipends of a number of technical officers who would be exempted from National Service in order to work under his direction. The new team was assembled, and a start was made as soon as the university had approved the proposals in the Faculty Board's report. The first to arrive was Mr Richard Weck, who had been working for the Welding Research Council since the previous February. Next, in January 1944 came Dr J. W. Roderick, Professor Baker's senior research assistant at Bristol, who gave up the lectureship at Leeds University which he had held during the war in order to take charge of the research team. He was made an assistant director of research in October 1945.

The completion of the temporary Structures Research Laboratory in the autumn of 1944 enabled work to be started in earnest. Some of Professor Baker's research apparatus was transferred from Bristol and more was made in the department workshop with funds supplied by the Research Council augmented by a small grant from the university. More technical officers[1] joined the team in 1945 and 1946 and all were put to work on the plastic theory of steel structures under Dr Roderick. Their main piece of equipment was a steel erection twelve feet high mounted on a concrete base which was christened 'the cathedral'. When the war ended the technical officers were allowed to register as research students, and by 1950 most of them had been admitted to the Ph.D. degree.

While Dr Roderick's team was at work on the plastic theory Professor Baker acceded to a request by the Admiralty to extend the scope of his researches in order to tackle a problem of immediate importance. The American all-welded Liberty ships and oil tankers, mass-produced to replace losses due to submarine and air attacks, were showing a tendency to break in half not only in rough weather but occasionally when in dock. This weakness was believed to be due to residual stresses

[1] Including Messrs B. G. Neal, M. R. Horne, and J. Heyman, all of whom played a key part in subsequent developments.

near the welded joints, and early in 1943 the Admiralty set up a Joint Committee with the Welding Research Council to investigate the matter. In 1945 the research was transferred to Cambridge and Mr Weck, who had already done some work on the subject, was taken from the structures team and put in charge of it. He was joined by Mrs C. F. Tipper, who had been working in the department with Professor G. I. Taylor during the war. At first they worked together, but after a time their interests diverged and while Mrs Tipper concentrated on the phenomena of brittle fracture, which she rightly suspected to be the cause of the Liberty ship disasters, Mr Weck resumed his study of the residual stresses in welded joints and problems of metal fatigue.

He started his research on the fatigue of welded structures in a small hut near Scroope House, using the method of resonance vibrations, but the noise led to many complaints and Professor Baker decided to move the research to another site. Early in 1946, at the suggestion of the Department of Scientific and Industrial Research, the Welding Research Council transformed itself into the British Welding Research Association and in this capacity purchased a property of about twenty-six acres at Little Abington, seven miles south-east of Cambridge. The estate included an old manor house and a number of disused army huts, one of which was fitted up as a research laboratory. An open-air testing ground was also established, and for the next four years the Little Abington site became in effect an outpost of the Engineering Department. Later B.W.R.A. built a fine range of laboratories and offices—all steel framed and designed on the plastic theory—and in 1956 Dr Weck (as he had become in 1948) left the department to become its director.

The original programme of Admiralty sponsored research on the structural behaviour of riveted and welded ships was completed in 1951. However, in 1954 a British-built tanker, the *World Concord*, sailing for Greek owners under the Panamanian flag, broke in two in the Irish Sea and Dr Tipper (a Doctor of Science since 1949) was commissioned to study and report on the disaster. She completed the task in 1959 and then resigned the readership to which she had been appointed ten years earlier and retired to the Lake District to write up the results of fifteen years' work on the phenomena and causes of brittle fracture.

By 1947 the investigations into the load-carrying capacity of rigidly-jointed single-storey steel-framed structures had proceeded far enough for a start to be made on the formulation of a design method based on the condition at collapse due to the formation of plastic hinges. Exhaustive series of tests had been carried out under static loading conditions on stanchions, beams, portals and welded connections, and many of the practical problems raised by the use of the plastic theory had been solved. In 1948 a clause was inserted in British Standard 449 permitting the plastic design of steel buildings. The method is simple and elegant, and when applied to single-storey frames achieves a saving of about 25 per cent in steel in comparison with traditional designs. It has been used successfully in many buildings, including the Structures Research Laboratory in Stage II of Baker Building.

Dr Roderick left the department in 1951 to become Challis Professor of Engineering and Dean of the Faculty in Sydney University, but work on the plastic theory continued. Having achieved success with single-storey buildings the team turned its attention to the study of multi-storey frames and pitched portals and the effect of high winds on tall structures. This problem proved to be much more intractable, but steady progress has been made and although a complete solution has not yet been found the method was used successfully in the design of the four-storey centre wing of Baker Building.[1] Work on the plastic theory reached a climax in 1956, when a three-day symposium held in the department was attended by over a hundred engineers and applied mathematicians from all over the world. In the same year a history of the investigation extending over twenty years was published by the Cambridge University Press under the title *The Steel Skeleton*, volume II, with Professor Baker, Dr Horne and Dr Heyman as joint authors.[2]

It was a logical extension of previous work on multi-storey buildings to make use of the strength of the concrete floors by tying them to the steel skeleton, and when Mr R. P. Johnson joined the staff as a lecturer in 1959 he began work on the plastic behaviour of these composite structures. As already described,[3] a composite design was accepted for the north wing of Baker Building.

[1] See p. 192.

[2] *The Steel Skeleton*, volume I, dealing mainly with the behaviour of steel structures in the elastic range, was published in 1954.

[3] See p. 193.

During the period covered by this review other members of the Structures and Materials groups have carried out successful research in such fields as the dynamic properties of steel and certain non-ferrous metals, pre-stressed concrete problems and the use of light alloy members in construction.

Professor Baker brought with him from Bristol a firm belief in the need for a limited number of university courses for graduates in engineering who have spent some years in industry. The object he had in mind was to give key men instruction in the latest technological developments made in their own subjects and so ensure that useful discoveries made in academic laboratories and elsewhere would be employed in industry with the minimum possible delay. Some years elapsed after his arrival in Cambridge before the staff could be spared or the money and equipment provided for this purpose, but in October 1951 six men with high qualifications and good industrial experience started work under Dr Weck on an eight-month post-graduate course on the Theory of Structures and Strength of Materials. Advanced instruction was given in matters of basic importance in the design of structures and students were encouraged to undertake an investigation into an appropriate problem under supervision. Additional graduate and assistant staff were paid for under the 1954 grant for higher technology and the course was put on a permanent basis. It was repeated each year until Dr Weck left the department, when it was taken over in succession by Dr Horne, Dr Heyman and Dr Biggs.

In 1964 the course was extensively revised, and since 1965 it has been described as an Advanced Course in Engineering Design Methods. Like its predecessor in recent years it will lead to a Certificate of Advanced Study in Engineering, one of the awards made by the university to candidates who have passed an approved examination. The new course is intended to help the young designer to develop his full potential and to show him how he can use analytical tools creatively. It is intended for men with at least one year's experience in design in addition to other practical experience who have a sound honours degree in engineering or equivalent qualifications. Each candidate will be expected to bring with him some difficult problem, of serious current concern to his firm, to serve as a focus for his own thinking and as an example for discussion by the course as a whole. Although the central theme is intended initially to be mechanical design, designers working in structural engineering may also be accepted. The first course started in October 1965.

Soil mechanics

Soil mechanics is concerned mainly with the engineering properties of soils and the prediction of their deformation when subjected to loads. The subject is obviously of particular importance to civil engineers, and theoretical investigations into some of the problems involved were carried out in various parts of the world during the seventeenth and eighteenth centuries. Interest lapsed for a time after 1850 but revived throughout Europe and the United States after the First World War. Little work was done on it in Britain, however, until the Building Research Station was established in 1933. Professor Baker introduced soil mechanics research into the department very soon after he assumed office. Mr K. H. Roscoe, who had graduated in 1937, was admitted as a Ph.D. candidate in October 1945 to research into the mechanical properties of clay and Mr A. A. Wells, now Professor of Structural Science in Queen's University, Belfast, started work on the mechanics of soil in relation to tillage in the Long Vacation of 1946. Mr Roscoe, however, resigned his research studentship in December 1946, when he was appointed demonstrator, and assumed responsibility for the soil mechanics teaching.

There was at that time little space available for research and the necessary equipment was virtually unobtainable, but an old store room in Scroope House was allocated and Roscoe converted it into a Soil Mechanics Laboratory. When the next academic year opened in October 1946 sufficient progress had been made for soil mechanics to be taught in this laboratory to candidates for the tripos. By this time Roscoe had also embarked on a programme of research into the shear behaviour of soils and had designed a new type of test equipment called the simple shear apparatus which was made in the departmental workshops. This proved to be the beginning of a partnership between the workshops and the Soil Mechanics Laboratory which over the years resulted in the development of a wide range of soil testing equipment of increasing capacity and accuracy. The work has had two main objectives—to devise more accurate test equipment and to produce apparatus versatile enough to enable a large number of different earth pressure problems to be investigated at all stress levels up to and beyond the point of failure.

A new Soil Mechanics laboratory was included in the first stage of Baker Building which opened in 1952[1] and in 1954 Mr Roscoe (now a University Lecturer) was joined by Messrs A. N. Schofield and C. P. Wroth.[2] These three, aided by research students, have succeeded in crystallizing into a single picture data on stress–strain characteristics of soils which had previously appeared to be unrelated. This was made possible mainly by using data obtained from simple shear apparatus and by working with soils under carefully controlled conditions. At the time of writing the soil mechanics group is developing the pattern of this picture and the theories derived from it with the ultimate object of explaining the observed behaviour of soils in the field.

In 1952 a new, but related, line of research in soil mechanics was initiated when Professor Baker decided to design by plastic methods a structure and its foundation as one unit and to test frames resting on economically possible foundations in real soils. Up to that time the experimental work of the structures group on steel frames had been carried out on foundations which were specially made to ensure rigidity but which were uneconomical for practical purposes. The work was carried out on a test bed on the B.W.R.A. estate at Abington. Some results of this work were presented at the 1956 Symposium on the Plastic Theory and were used a year later in the design of a large single-storey test house for B.W.R.A. Following this field work versatile laboratory test equipment has been developed in which a large number of different earth pressure problems can be investigated under controlled conditions and in which X-ray and radio-isotope techniques can be used to determine the strain pattern throughout the soil mass. One of the group's aims is to predict these patterns from the stress–strain theories and some success in this direction has already been achieved in preliminary tests.

[1] It was doubled in size when the North Wing was completed in 1964.

[2] Both later admitted to the Ph.D. degree and appointed to university lectureships.

By the end of 1965 the Soil Mechanics group numbered twenty-six in all. Of the fifteen former research students four were on the staff of the department and eight were, or had been, heads of other soil mechanics laboratories, three of them as full professors.

Fluid Mechanics

In the sub-department of Aeronautics teaching and research had proceeded hand in hand since the foundation of the chair in 1919, and reference has already been made to the work of Professor Melvill Jones between the wars. In 1929 he opened up a new field of research when, in a lecture to the Royal Aeronautical Society, he drew attention to the fact that about two-thirds of the propulsive power of a typical transport aircraft was expended in the generation of unnecessary eddies. From that time a great deal of his research was related to the reduction of drag, and with this end in view he directed a series of flight experiments which were carried out from Duxford aerodrome. In the first Wright Brothers Lecture in 1937 he drew attention to the factors controlling transition of the boundary layer on a wing. After the war he directed the work of a team whose attention was concentrated on the reduction of drag by distributed suction applied through the wing surface. From 1948 to 1950 a series of flight experiments was performed, using a small aerofoil mounted under the fuselage of an Anson aircraft.

When Professor Mair succeeded to the chair in 1952 work on boundary layer control was continued with wind tunnel experiments on suction applied to the turbulent boundary layer. At the same time a method was developed by J. H. Preston[1] for measuring skin friction in turbulent boundary layers on solid surfaces. In 1954 flight experiments were resumed on suction for low drag, using a Vampire aircraft which had been modified at the R.A.E., and this work was brought to a successful conclusion by 1957. However, the proposal for the construction of a complete laminar-flow research aircraft was not accepted by the Ministry of Supply, which had supported the work up to this time, and the research group turned its interest to the use of suction for high lift. When the new laboratories in the south wing came into use in 1958 facilities for experiment were greatly improved and a

[1] Now Professor of Fluid Mechanics at Liverpool University.

number of new research topics were undertaken. In 1960 a high-speed wind tunnel of the blow-down type was commissioned and has since been used for experiments on ventilated walls, the flow round a jet issuing into a supersonic stream and three-dimensional shock-wave boundary layer interactions.

Flight experiments on the use of suction for high lift were started in 1961 using an Auster aircraft modified by the provision of new wings and a suction unit driven by a small gas turbine. In experiments which are still continuing the lift coefficient has been raised, without the use of flaps, from 1·6 to about 5·5. In connection with these experiments a method has been developed for calculating the turbulent boundary layer with suction or blowing, and the performance aspects of take-off and landing at high lift coefficients have been analysed.

In 1963 the equipment of the laboratory was completed with the commissioning of the large return circuit wind tunnel. This has been used for research on a number of problems, including unsteady flow in cavities, the lifting properties of slender wings and boundary layer control by suction. Other research in progress at the time of writing includes spoiler controls, drag of bluff bodies, three-dimensional boundary layers and the effect of air injection into a boundary layer.

Experimental research on hydraulic problems has been carried out in the department since Mr A. M. Binnie joined the staff as a lecturer in 1945.[1] Much of it was originally prompted by the increasing use of trumpet-shaped overflows for removing surplus water from reservoirs, as discharge through these devices can be greatly reduced by whirlpools. A study was therefore made of the behaviour of a revolving flow of water through a nozzle, both from an open tank under the action of gravity and from a closed tank under pressure. More recently revolving flow in straight pipes and in bends has been examined and hydraulic jumps akin to those found in rivers and open channels have been discovered.

Other matters which have been investigated include the use of air bottles to protect pipe lines from surges and the detection of turbulence in pipe flow by a double refraction method, and self-incited oscillations

[1] He was promoted to a readership in 1954.

have been studied in a circular tank with a fountain and in a channel with corrugated walls. The time of passage of small spheres of different size and density injected in turn into a horizontal water pipe has been measured and statistical theory applied to the observations.

Apparatus has also been developed to produce in an open channel a stream of water, uniform in depth and velocity, in which ship models could be held. In this way naval architects might be provided with facilities analogous to wind tunnels. Considerable success was achieved, and large plant has been constructed on the same lines elsewhere. Later a refinement was tested which employed slotted walls surrounded by a large chamber containing almost stationary water. This device allows proportionately larger models to be used.

The Suez affair of 1956 and the subsequent re-routing of shipping round the Cape led to a shortage of tankers, and as a possible remedy Professor Hawthorne formed a group with Mr J. C. S. Shaw and Sir Geoffrey Taylor to develop the idea of using flexible plastic containers for the transport of oil fuel. This led to the study of a series of problems in fluid mechanics, and experimental and theoretical work was started on a project under the name of 'Dracone' which was sponsored by the National Research Development Corporation. Much of the work has been concerned with the flow of liquids through and around flexible tubes and with the stability of towed flexible objects and their behaviour in waves. In the early stages the work was centred mainly in the department, but a group in Southampton under N.R.D.C. sponsorship later took over the development of these Dracones, several of which are now in commercial use in various parts of the world.

Electrical engineering

The appointment of E. B. Moullin as Professor of Electrical Engineering in 1945 did not lead immediately to the expansion of research in this field as, although he had done some very distinguished work himself, his interests in the department were directed mainly towards undergraduate teaching. However, the Governing Body of Trinity College decided in the same year to elect three professional engineers to fellowships on the understanding that they would be appointed to university lectureships in the department of Engineering. Among those elected was Mr C. W. Oatley, head of the Army's Radar Research and Development Establishment. His chief interest was electronics, and on joining the staff of the department he initiated an extensive programme of research in that subject. In 1948 he was joined by Mr J. G. Yates, who had worked with him during the war, and together they built up and supervised a team of research students, some of whom afterwards became members of the teaching staff. Financial support was obtained from the Admiralty, and a research contract signed in 1951 was still in operation in 1965.

For ten years the work of the Electronics group was concentrated mainly on the development of the scanning electron microscope and on problems related to the motion of electrons in micro-wave valves. The death of Mr Yates in 1956 was a sad blow to the department, but the appointment in 1958 of Mr A. H. W. Beck, an expert on valve techniques, marine radar and other branches of electronics, brought in new ideas and resulted in a considerable extension of the field covered by the group. Under his direction research on high-current density cathodes was started with the object of improving the high frequency and high-power performance of the microwave valves used in communication links and radar. In the course of this investigation several types of cathodes have been examined in the scanning electron microscope and much new information has been obtained on the mechanism of electron emission. The research continues.

When Mr Oatley succeeded to the professorship in 1960 he encouraged an increase in research in heavy-current electrical engineering as well as in electronics. As a result of this policy a flourishing group is now engaged on the study of various aspects of the automatic control of electrical machines, and detailed studies of non-linear processes in such devices have been undertaken using analogue computers. Work has also been started on high-current mercury arc discharges. The work on scanning electron microscopy has led to the development of a series of unconventional electron probe instruments. Among these has been the magnetically scanned electron diffractometer, a new instrument likely to supersede the traditional photographic methods in electron diffraction. Use of these instruments has led to the discovery of a new state of matter, the amorphous magnetic state, and to new knowledge of the nucleation of metal films on the atomic scale.

The electronic laboratories on the top floor of the north wing of Baker Building came into use in October 1964, and much of the research formerly carried out in the Inglis laboratories was transferred to them. The heavy equipment was moved to temporary quarters on the Pitt Press site, where it will remain until the completion of 'Inglis A' in November 1966 makes it possible to bring all the teaching and research in electrical engineering together again.

Applied thermodynamics

Mention has already been made of the research into problems of combustion and applied thermodynamics which was carried out in the department while Bertram Hopkinson was professor. Interest in the subject was maintained during Inglis's time by Mr A. L. Bird, the Hopkinson Lecturer in Thermodynamics, who made valuable contributions to the study of combustion and scavenging problems in the internal combustion engine.

In 1946 Mr H. G. Rhoden, who had joined the staff in January 1939 from the Turbine Department of Metropolitan Vickers, proposed to the Ministry of Aircraft Production that a wind tunnel should be constructed for research on the flow through cascades of blades in axial compressors, which were a major component of some types of jet engines developed for aircraft propulsion during the war. The proposal was discussed with Dr W. R. Hawthorne, then Deputy Director of Engine Research in the Ministry, and a programme of work under Mr Rhoden's direction was approved. This proved to be the starting-point of a substantial programme of research on the aerodynamics of compressors and turbines which has been supported over the years by the government department, now renamed the Ministry of Aviation.

In 1951 Professor Hawthorne was elected to the newly established chair of Applied Thermodynamics. During the war he had been responsible for the design of the combustion chambers of the first Whittle jet engine to fly and had then been in charge of the Gas Turbine Division of the Royal Aircraft Establishment at Pyestock (now the National Gas

Turbine Establishment). After the war he had continued to do research at the Massachusetts Institute of Technology on gas dynamics, combustion and the aerodynamics of compressors and turbines. In 1951 there were growing signs that the development of axial flow compressors for jet engines was becoming increasingly difficult. Only major aero-engine manufacturers could deploy the effort, sometimes extending over years, which was required to get the compressor for a new engine performing satisfactorily. In one or two cases these efforts failed and the difficulties were contributory factors to the withdrawal of several firms from the aero-engine business in the early fifties.

On his appointment Professor Hawthorne began to build up a greatly increased activity in axial compressor research with the intention of improving design methods. First efforts were concentrated on his actuator disk method which promised to account more satisfactorily for the three-dimensional flows. J. H. Horlock[1] was the first research student to use a compressor presented by Rolls Royce in 1952 which was installed in the old smithy at the south-west corner of the Inglis building. He showed that the disk method was substantially better than earlier radial equilibrium methods, and several calculations based on the method were performed for industry, including one for an 8-stage compressor for an American firm. Work was continued by other research students using annular cascades and the Rolls Royce compressor. A paper reviewing the work was presented to the Institution of Mechanical Engineers by Professor Hawthorne and Horlock in 1962 and was awarded a James Clayton Prize. The extension of the research to compressible flows was made possible by permission to use the National Gas Turbine Establishment's facilities for some experiments. The actuator disk theory predicts that under certain conditions standing waves may be found in the flow in turbo-machines. The existence of these waves, which were not predicted by earlier theories, was demonstrated experimentally for the first time in 1965. Throughout this period computers have been increasingly used. At Professor Hawthorne's instigation a project on the computation of flow in compressors centred at the National Gas Turbine Establishment was held in 1965 which was attended by research workers from universities and firms. Further

[1] Now Professor of Mechanical Engineering at Liverpool University.

development of such design methods continues and a post-graduate course in the computer aided design of turbo-machinery is now being offered.

In order to develop satisfactory design methods many details of the flow in turbo-machinery need to be well understood. One such phenomenon is the secondary flow which occurs on the walls or in a bent duct. Work on secondary flow has continued steadily since 1951 and numerous papers have been published. The work has led to new developments in the general theory of three-dimensional flows which is the subject of recent post-graduate lectures by Professor Hawthorne.

Another phenomenon of importance is the vibration of blades in turbo-machinery. Such vibrations may be the result of flutter. In 1954 D. S. Whitehead began a programme of work on aero-elastic vibrations of blades which is still continuing. Several reports of this work have been published and computer programmes have been prepared for industry. Blade vibration may also occur during stalling and a study of stall propagation in axial compressors and cascades was started by M. D. Wood in 1952. Hot wire anemometers have been used extensively during this work.

The work on stationary cascades started by H. G. Rhoden has led to the building of two new cascade tunnels, and in 1964 he was responsible for the building of a large, well-instrumented rotating cascade wind tunnel.[1]

In 1947 a programme of research on the flow of gas and oil past piston rings in internal combustion engines was started by Dr P. de K. Dykes, whose work was supported by the Motor Industry Research Association. Using sophisticated techniques of measurement he explained the excessive gas blow-by at high speeds and developed a new type of piston ring which is now universally used in racing-car engines. The results of the work on oil flow have been used successfully to reduce the oil consumption of automobiles, rail locomotives and ships. Papers were read on both aspects to the Institution of Mechanical Engineers,

[1] During this period over forty papers on flow in turbo machinery have been submitted to the Aeronautical Research Council or published in technical journals. In 1958 *Axial Flow Compressors* by J. H. Horlock was published by Butterworths. Professor Hawthorne was editor of and contributor to *Aerodynamics of Turbines and Compressors* and *Design and Performance of Gas Turbine Power Plants*, Volumes X and XI of *High Speed Aerodynamics and Jet Propulsion*, published by the Oxford University Press in 1964 and 1960 respectively.

and resulted in the award of the James Clayton Prize, the Crompton-Lanchester Medal, and the T. Bernard Hall Prize.

In 1948 the study of heat transfer to liquid metals was started with the support of the Atomic Energy Authority, which was interested in the possibility of using liquid metals for cooling fast reactors. The research soon turned to the problem of pumping the metals by electromagnetic means and the use of electromagnetic devices for measuring flow rates. One of the important early results was the discovery by Dr W. Murgatroyd[1] that the condition for transition from laminar to turbulent flow of liquid metals in channels is clearly related to the magnetic field present. Dr J. A. Shercliff[2] continued the work on the behaviour of liquids in channels and succeeded in finding results by which pressure losses in electromagnetic flowmeters could be predicted. Under Dr Shercliff, the scope of the work on liquid metals was extended to include more general investigation in magnetohydrodynamics, a notable success being the first convincing demonstration of Alfven waves in liquids. The scope of the work was further extended by the construction of a 5 in. diameter shock tube in which the behaviour of electrically conducting gases or plasmas can be examined. Magnetohydrodynamic studies of both liquid metals and plasmas are being continued. The Atomic Energy Authority has also supported research on heat transfer to non-Newtonian fluids which might be used in organic liquid moderated reactors and to gases containing clouds of solid particles.

Work on the behaviour of heated evaporated liquids in boiler tubes was started by Mr R. W. Haywood in 1949. The Water-tube Boilermakers' Association then sponsored the construction of a full scale experimental rig which was erected on a site off the Madingley Road about three miles from Cambridge. The results of the work on this apparatus provided much fundamental knowledge of value to boiler designers, and a paper describing the work was awarded the Thomas Lowe Gray Prize by the Institution of Mechanical Engineers in 1961.

Between 1950 and 1954 D. B. Spalding,[3] first as a research student and then as a member of the teaching staff, was responsible for research on combustion. He developed a new method of measuring flame speeds

[1] Now Professor of Nuclear Engineering, Queen Mary College, London.
[2] Now Professor of Engineering Science, University of Warwick.
[3] Now Professor at Imperial College, London.

and produced theories of heat and mass transfer with special applications to combustion. Other workers in the Heat group studied heat transfer by natural convection to fluids of variable viscosity and the flow of fluids of non-uniform density over bodies of various shapes.

Mechanics of machines and control engineering

Problems of mechanical vibration have been studied in the department since the days of Ewing and have occupied a central place in the research activities of the Mechanics group which followed the post-war expansion of the teaching staff. The photoelastic investigation of the stresses in gear teeth under dynamic conditions was started in 1946 by Mr E. K. Frankl and led to the discovery that the observed stresses were related to the vibration characteristics of the complete gear. This conclusion led to a detailed study by various members of the group of the dynamic behaviour of general systems which was further stimulated by the election of Professor D. C. Johnson to the newly created chair of Mechanics in 1962.

The appointment of Dr S. A. Tobias[1] to an assistant directorship of research in 1955 opened up the new field of machine tool vibration. The early work was mainly confined to milling machines, but at the time of writing interest is focused on the vibration characteristics of gear cutting machines and their influence on the quality of the gears which are produced.

During the decade since 1954 Dr K. L. Johnson has conducted a continuous programme of research into the mechanics of surface contact, with particular reference to the stresses and deformation which occurs at highly loaded rolling contacts such as ball bearings and gear teeth. This work has proved to be complementary to the famous study of the physical nature of the frictional phenomena by Dr F. P. Bowden and Dr D. Tabor in the Cavendish Laboratory.

Work on mechanical control systems was started by Mr R. H. Macmillan[2] in 1947 and was continued by Dr G. D. S. MacLellan,[3] Mr D. B. Welbourn and Dr P. E. W. Grensted. The field was greatly extended

[1] Now Professor of Mechanical Engineering at Birmingham University.

[2] Later Professor of Mechanical Engineering at the University College of Swansea, now Director of the Motor Industry Research Association.

[3] Now Rankine Professor of Mechanical Engineering at Glasgow University.

when Mr J. F. Coales was appointed an Assistant Director of Research in 1952. Mr Coales had been Research Director for Elliott Brothers (now Elliott Automation) and had built up large laboratories working mainly in the fields of electronic computing, instrumentation and control. With the help of members already working in the Mechanics group he drew up a long-term programme of research into the analysis, optimization and synthesis of non-linear control systems, and financial support was obtained from the Ministry of Supply. It soon became necessary to appoint additional staff and the Control Engineering group eventually hived off from the parent body. The scope of its activities widened and it became the first group in the United Kingdom to attack the problem of analysing and optimizing non-linear control systems with random inputs. It soon achieved an international reputation, particularly for its work on the use of functionals in systems analysis and on non-linear filters in the development of predictive control.

The first research student joined the Control group in 1953 and the following year saw the start of a post-graduate course in Control Engineering which, like the Structures Course already referred to, led to a Certificate of Advanced Study in Engineering. This course has been very successful and has continued to attract highly qualified students from industry and the Services. New methods of teaching control theory and systems design have been developed with particular emphasis on statistical methods and design analysis.

In 1961 the National Research and Development Corporation made available a grant of £50,000 for the purchase of an analogue computer and the National Development Corporation and Associated Electrical Industries Ltd. made a gift to the department of an Elliott 405 digital computer. The old gatehouse at Madingley Road was reconditioned as a computing centre and the machines were installed in it.

By this time the work of the group had reached a stage when it could be extended to the application of control theory to the more practical but far more difficult problems associated with the optimization of industrial processes. Increased financial support by D.S.I.R. made it possible to obtain the services of Dr H. H. Rosenbrock,[1] a specialist in

[1] Now Professor of Control Engineering at the Manchester College of Science and Technology.

the optimization of systems design by computational methods. A programme of work was drawn up which aimed at finding methods for the adaptive control of complex industrial plant, and as the group by now consisted of seven members of staff and twenty research students it became possible to direct a multipronged attack on this important problem. In May 1964 another grant, totalling £275,500, was sought from D.S.I.R. for the replacement of the Madingley Road computers by more powerful and versatile machines and to further a long term programme of research in close collaboration with industry. Confirmation of this grant was made to the department in July 1965 by the Science Research Council (successor to D.S.I.R.)

TAILPIECE

Over one hundred and eighty years have passed since the first lectures on engineering were given in Cambridge, and the subject which was originally no more than the hobby of gifted and eccentric professors is now universally acknowledged as one of the major concerns of a modern university. The department founded by James Stuart in 1875 now has a graduate staff of ninety and nearly a thousand students. The wooden workshop and borrowed lecture-room have given way to a complex of buildings whose equipment will bear comparison with that of any Engineering School in the world. Cambridge educated engineers hold many of the key positions in British industry, and the research which is carried out in the laboratories plays a vital part in the advance of the technology on which the development of our modern civilization so largely depends. With the retirement of Professor Sir John Baker in 1968 another chapter in the long history of the department of Engineering will come to a close, but before he goes the addition of new professorships and the completion of the first section of the new Inglis Building will pave the way for further advances in both teaching and research. There is no reason to doubt that the future of the department will be as interesting and exciting as its past.

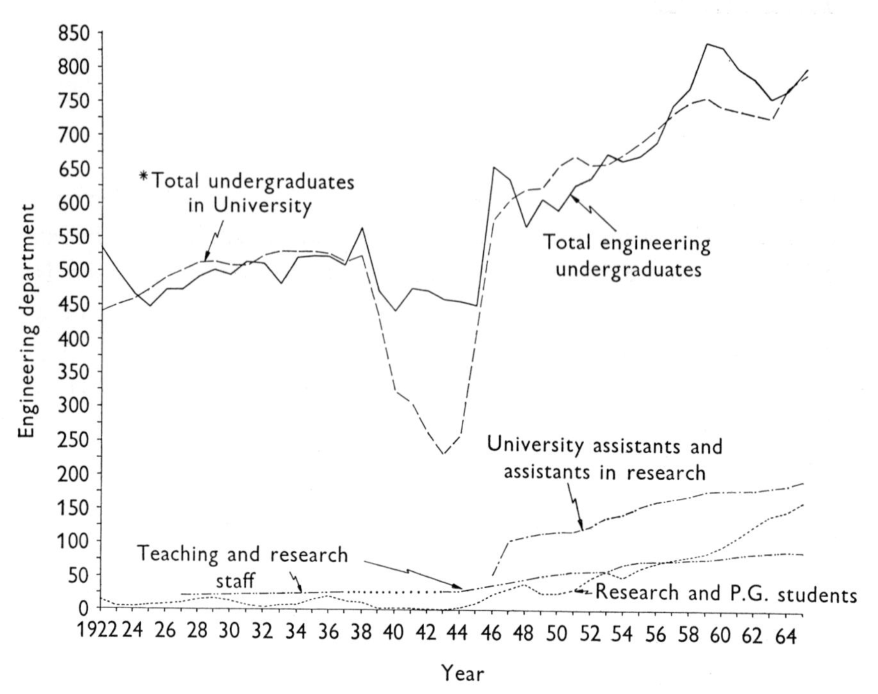

Graph showing the growth of the engineering department, 1922–65.
* Total number of students in the university. $\times \frac{1}{10}$.

APPENDIX
THE CAMBRIDGE UNIVERSITY ENGINEERS ASSOCIATION

The forerunner of the Cambridge University Engineers Association was the Engineering Society for undergraduates founded under the auspices of Professor Ewing in 1902. One of the founder members of the Society was A. D. Browne of Queens', who succeeded E. D. A. Herbert as its president in 1924. The Society's main object was to encourage undergraduates to interest themselves in the practical problems of the engineering industry and relate them to the lectures given for the tripos. J. B. Peace was the first treasurer and F. J. Dykes and C. E. Inglis were among its early supporters. Under this energetic leadership the Society prospered, and in 1924 arrangements were made for the institution of life membership, which carried with it an entitlement to all the privileges of full membership, including delivery of the C.U.E.S. *Journal*, whose first number had appeared in December 1923. Three years later the Engineering Society joined forces with the Aeronautical Society founded by Professor Melvill Jones's students. The list of life members published in the 1929 *Journal* contained 134 names.

In spite of the opportunity for life membership offered by the Engineering Society Mr W. N. C. van Grutten of the University Appointments Board felt that there was a real need for an organization which would keep Cambridge engineers in touch with the department and with each other. This would apply not only to those who were just graduating but also to those who had distinguished themselves in the profession and whose achievements would be an inspiration to young men starting on their careers. In 1928 therefore he proposed to Professor Inglis that a Cambridge University Engineers Association should be founded on the lines of the Association of Officers of the Division in which he served in the First World War. Inglis keenly welcomed the idea, and in December 1928 a small number of senior Cambridge engineers who were interested in the proposal met in the offices of the consulting engineers firm of Sir Alexander Gibb and Partners at the invitation of Mr Rustat Blake of Pembroke, senior partner in the firm. Those present formed themselves into an Executive Committee, and a letter was written to many graduates in industry inviting them to co-operate in establishing an Association by joining its first General Committee. The response was enthusiastic, and at a meeting on 27 May 1929 the resolutions of the Executive Committee were unanimously approved and the Association formally constituted. Its objects were defined as follows:

(*a*) To keep all past and present members of the University who are or have been associated with the profession of Engineering in touch with each other and with the University Engineering School.

(*b*) To compile and keep up to date as complete a record as possible of the careers of all Cambridge engineers, and to publish annually for issue to its members a Register giving the names and addresses of Cambridge engineers.

(*c*) To organise such reunions of its members as may be considered of value to the Association.

(*d*) To facilitate united action on any matter concerning the welfare of Cambridge engineers.[1]

Sir Charles Parsons was elected president and the vice-presidents were Sir Alfred Booth of the Cunard Steamship Company, Sir Alfred Ewing, lately retired from the Vice-Chancellorship of Edinburgh University, and Lord Melchett of Imperial Chemical Industries Ltd. Professor Inglis became chairman of the Executive Committee and Mr W. N. C. van Grutten honorary secretary and treasurer. A. D. Browne and J. W. Landon served on the General Committee which had a total membership of thirty-two, and included E. D. A. Herbert, who had been president of the C.U.E.S., Harry Ricardo and many other engineers of outstanding ability and influence.

The early years of the Association were occupied mainly in tracking down Cambridge graduates by searching old records. In this work valuable assistance was given to van Grutten by Charles Macneil of Jesus, who dealt with the Engineering Institutions, and by H. A. Roberts, who allowed the files of the University Appointments Board to be examined for this purpose. Those who were located were invited to obtain life membership by paying a single subscription of three

[1] Cambridge University Engineering and Aeronautical Society's *Journal*, 1930, p. 16.

guineas. There was a good response to the appeal and among those who joined were men of eminence in all branches of the profession. In addition there was a steady influx of men who went down after graduation, particularly those who had passed through the hands of Mr van Grutten. Members were issued annually with a *Year Book* which contained the name, college and date of degree of every graduate who had been located and articles reprinted from the *Journal* of the undergraduates' Engineering and Aeronautical Society. The 1938 *Year Book*, which proved to be the last of the series, contained about 2,400 names. Afternoon reunions followed by dinner in a college were held in 1933 and 1937.

No meetings were held during the war, but in 1947, at Professor Baker's suggestion, it was agreed that the Register of Members should be brought up to date as a first step towards the revival of Association activities. At a General Meeting held on 8 July 1950 Professor Inglis was elected president, Professor Baker chairman and the secretary of the department, A. H. Chapman, honorary secretary and treasurer. The first post-war reunion was held on 14 July 1951.

Sir Charles Inglis died in April 1952 and was succeeded as president by Sir Edward D. A. Herbert, who had been a member of the Executive Committee since it was formed in 1929. In May 1953 His Royal Highness the Duke of Edinburgh, who had opened the first block of Baker Building in the previous November, consented to become Patron of the Association. A revised constitution, which had been submitted to members in draft form in 1952, was approved at the Annual General Meeting held on 11 July 1953. The officers of the Association were to consist of a president, a vice-president (normally the Head of the Department) and an honorary secretary and treasurer. The vice-president was to be chairman of an Executive Committee consisting of six resident and six non-resident members. The Association was to be open to all Cambridge graduates who were, or had been, associated with the profession of engineering. In the following year the privileges of membership were extended to all graduates of the department for two years after going down without payment of a subscription.

At the Annual General Meeting in 1956 it was agreed that subsidiary branches should be established outside Cambridge to keep members living or working in the same area in touch with each other and to foster the general aims of the Association. The first provincial branch was inaugurated in the West Midlands in 1957 by Mr J. W. Warner, a member of the Executive Committee, and was quickly followed by a branch in the East Midlands under the chairmanship of Sir Edward Herbert. The London area followed suit in 1962 with Mr P. F. Grove as chairman and a flourishing branch was established in South Africa after Professor Baker's visit in the same year. The popularity and usefulness of the local branches is confirmed by a steady growth in membership and in the range of their activities.

Sir Edward Herbert died on 28 April 1963. He had given the Association active and generous support ever since its foundation, and his death was a great loss. He was succeeded as president by Lord Brabazon of Tara, but he too died within a year of his election and was succeeded in 1965 by Lord Nelson of Stafford.

A decision which it is hoped will have far-reaching consequences was taken in 1963 when it was decided that the reunion in the laboratories would in future be preceded by a technical conference at which papers on some subject of topical interest to the profession would be presented and discussed. The subject chosen for the first conference, which was held on Saturday 28 September 1963, was 'Industry and the Universities' and the introductory papers were presented by Lord Nelson of Stafford and Professor Baker. About two hundred members attended and there was a lively discussion which was published in the Report of the Conference. At the Annual General Meeting which followed it was decided to set up a working party to decide on the subject for the 1964 meeting and prepare papers for circulation.

The working party was chaired by Mr R. L. Fortescue, a lecturer in the department with considerable experience in industry, and his most active helper was Mr J. D. Brown, who had recently retired from the office of Engineering Director of a Division of Imperial Chemical Industries, Ltd. The subject chosen for the second conference, which was held on 26 Spetember, was 'The Training of Graduate Engineers'. It aroused much interest, and it is to be hoped that some of the views expressed in the discussion will carry weight not only in the firms in which members of the Association are working but in the councils of the professional institutions themselves.

The 1965 and 1966 conferences are to be devoted to a wide discussion of the kind of work which Cambridge men are called upon to undertake in engineering. The object will be to decide in what respects changes and progress seem most important, both in education and in the engineering profession as a whole, if engineering is to flourish in this country. The 1965 conference was held on 2 October and the discussion was opened by Viscount Caldecote, who was a lecturer on the staff of the department from 1948 to 1955, a director of the English Electrical Company and Deputy Managing Director of the British Aircraft Corporation (G.W.), Ltd.

At the time of writing the Association has over four thousand members, many of whom occupy key positions in industry, and there is ground for believing that these annual conferences will result not only in the production of new ideas but also in action which will be of positive benefit to the national economy.

BIBLIOGRAPHY AND NOTES

GENERAL

ASHBY, SIR ERIC. *Technology and the Academics*. Macmillan, 1958.

BALL, W. W. ROUSE. *A History of the Study of Mathematics at Cambridge*. Cambridge University Press, 1889.
[Main authority for the history of the development of the study of mathematics and applied science between 1600 and 1858.]

Cambridge University Calendar from 1802. Now entitled *Annual Register of the University of Cambridge*. Cambridge University Press.

Cambridge University Engineering Department. *Minute Books of Faculty Boards and Degree Committees*.

Cambridge University Reporter from 1870. Cambridge University Press.
[Contains Reports to the University by the Council of the Senate, General Board, Faculty Boards, Syndicates, etc. Lecture programmes and verbatim reports of discussions are printed in full. No page references are given in the text as the items referred to can be identified by their dates in the Index in the bound volumes.]

CARTTER, A. M. (ed.) *American Universities & Colleges*. American Council on Education, 1964.

Commonwealth Universities Yearbook. Association of Commonwealth Universities. Various dates.

Dictionary of National Biography. Smith, Elder and Co., 1885–1950.

International Handbook of Universities. International Association of the Universities, 1959.

ROBERTS, S. C. *Introduction to Cambridge*. Cambridge University Press, 1953.

Statutes and Ordinances of the University of Cambridge. Cambridge University Press.

STERLAND, E. G. Early History of the Teaching of Engineering in Cambridge. Typescript. Cambridge University Engineering Department Archives.

TANNER, J. R. (ed.) *Historical Register of the University of Cambridge*. To 1910 and Supplements 1911–20, 1921–30. Cambridge University Press.

TILLYARD, A. I. *A History of University Reform*. W. Heffer, 1913.

University Archives.
[Papers and newspaper cuttings, especially those in the volumes entitled *Professor of Mechanism and Engineering*.]

VENN, J. A. Oxford and Cambridge Matriculations, 1544–1906. *Oxford and Cambridge Review*, no. 3, 48 (1908).

VENN, JOHN and VENN, J. A. *Alumni Cantabrigienses.* Cambridge University Press, 1922.

Victoria History of the Counties of England. Vol. 3. *Cambridge and the Isle of Ely.* Oxford University Press, 1959. Section on the history of the University by the editor, Dr J. P. C. Roach.

[Main source for the University background to the history of the Department over the whole period covered.]

WINSTANLEY, D. A. *Unreformed Cambridge.* Cambridge University Press, 1935.

WINSTANLEY, D. A. *Early Victorian Cambridge.* Cambridge University Press, 1940.

WINSTANLEY, D. A. *Later Victorian Cambridge.* Cambridge University Press, 1947.

[Main authorities for the detailed history of the University from the mid-eighteenth century to 1882.]

CHAPTER I. THE UNIVERSITY BACKGROUND TO THE STUDY OF ENGINEERING

BUTTERFIELD, H. *Origins of Modern Science.* George Bell, 1958.

CROMBIE, A. C. *Augustine to Galileo.* Mercury Books, 1961.

GUNNING, H. *Reminiscences of the University, Town and County of Cambridge from the year 1780.* George Bell, 1854.

LATHAM, H. *On the Establishment in Cambridge of a School of Practical Science.* Cambridge University Press, 1859.

Report of H.M. Commissioners appointed to Inquire into The State, Discipline, Studies, and Revenues of the University and Colleges of Cambridge, 1852. *Parliamentary Paper, 1852–53* [*1559*] *XLIV, 1.*

Report of the Cambridge University Commissioners, 1867. *Parliamentary Paper, 1861* [*2852*] *XX, 53.*

Third Report of the Royal Commission on Scientific Instruction and the Advancement of Science, 1873. *Parliamentary Paper, 1873, C. 868, XXVIII, 637.*

WHEWELL, W. *Of a Liberal Education.* 1845–52.

WHITESIDE, D. T. Isaac Newton: birth of a mathematician. *Notes and Records of the Royal Society,* **19** (1) (June 1964), 53.

WHITESIDE, D. T. Newton's early thoughts on Planetary Motion: a fresh look. *Br. J. Hist. Sci.* **2**, Part II(6) (Dec. 1964), 117.

CHAPTER 2. PROFESSORIAL LECTURES ON APPLIED SCIENCE AND ENGINEERING, 1783–1875

Isaac Milner

ATWOOD, G. *A Description of The Experiments intended to illustrate a Course of Lectures on the Principles of Natural Philosophy*..., *Cambridge*. London, 1776.

GUNNING, H. *Reminiscences of the University, Town and County of Cambridge from the year 1780*. George Bell, 1854.

MILNER, I. *A Plan of a Course of Experimental Lectures*. Cambridge, University Printer, 1780.

MILNER, M. *Life of Isaac Milner, Dean of Carlisle*. J. W. Parker and J. and J. J. Deighton, 1842.

SOCIUS. *Facetiae Cantabrigiensis*. William Cole, 1825.

William Farish

FARISH, W. *A plan of a Course of Lectures on Arts & Manufactures*,.. Cambridge, University Printer, 1821.

FARISH, W. Obituary. *Christian Observer*, 611–13, 674–7, 737–41. 1837.

GUNNING, H. *Reminiscences of The University, Town and County of Cambridge from the year 1780*. George Bell, 1854.

Magdalene College Magazine, no. 86 (1955).

Magdalene College Records.

SOCIUS. *Facetiae Cantabrigiensis*. William Cole, 1825.

WORDSWORTH, C. *Scholae Academicae; some account of the Studies at the English universities in the 18th Century*. Cambridge University Press, 1877.

George Biddell Airy

AIRY, G. B. *Syllabus of Course of Experimental Lectures*. Cambridge, University Printer, 1826.

AIRY, G. B. *Autobiography*. Cambridge University Press, 1896.

BABBAGE, C. *Reflections on the Decline of Science in England*. B. Fellowes, 1830.

WOLLASTON, F. J. H. *A Plan of a Course of Chemical Lectures*. Cambridge, University Printer, 1794.

WOODHOUSE, R. *Syllabus of Lectures on Experimental Philosophy*. 1819.

Robert Willis

DOUGLAS, Mrs STAIR. *The Life & Selections from the Correspondence of William Whewell, D.D.* C. Kegan Paul, 1881.

Memoirs of Deceased Members: Robert Willis. *Minut. Proc. Instn civ. Engrs*, **41**, Part III (1874–5), 206.

Obituary Notice, *Cambridge Chronicle* (6 March 1875).

Report of the Commissioners appointed to Inquire into the Application of Iron to Railway Structures, 1849. *Parliamentary Paper, 1849* [*1123*] *XXIX.*

SHIPLEY, A. E. '*J*', *a memoir of John Willis Clark*. Smith, Elder, 1913.

WILLIS, R. *Syllabus of a Course of experimental Lectures on the principles of Mechanism.* Cambridge, University Printer, 1841.

WILLIS, R. *A System of Apparatus for the use of Lecturers and Experimenters in Mechanical Philosophy.* Weale, 1851.

WILLIS, R. *Principles of Mechanism.* Parker, 1841; 2nd ed. 1873.

WILLIS, R. Scrapbook. *Cambridge University Engineering Library.*

CHAPTER 3. JAMES STUART

ARMYTAGE, W. H. G. *Civic Universities.* Benn, 1953.

BALLOT, H. Hale. *University College London, 1826–1926.* University of London Press, 1929.

BROWNE, G. F. *Recollections of a Bishop.* Smith, Elder, 1915.

CHORLEY, KATHARINE. *Manchester made them.* Faber and Faber, 1928.

Departmental Record of Undergraduate Entries. *Cambridge University Engineering Department Archives.*

GREIG, J. John Hopkinson, 1849–1898. *Engineering, Lond.*, **169** (13 and 20 Jan. 1950), 34–6 and 62–4.

HEARNSHAW, F. J. C. *History of King's College, London, 1828–1928.* G. G. Harrap, 1929.

HOPKINSON, EVELYN. *The Story of a Mid Victorian Girl.* Cambridge University Press, 1928.

HOPKINSON, MARY and EWING, LADY. *John and Alice Hopkinson, 1824–1910.* Farmer and Sons, *c.* 1911.

Obituary notice of Sir A. B. W. Kennedy. *Minut. Proc. Instn civ. Engrs*, **227**, Part I (1928–9), 269.

STUART, J. *Reminiscences.* Chiswick Press, 1911.

STUART, J. Letters to his Mother. MS. Cambridge University Engineering Department Archives.

CHAPTER 4. J. A. EWING

EWING, J. A. *The University Training of Engineers; an introductory lecture.* Cambridge University Press, 1891.

EWING, J. A. *The Steam and other Heat Engines.* Cambridge University Press 1894.

EWING, J. A. Engineering School in its early Days. *Camb. Univ. Engng Soc. J.* **1**(3) (1923), 4.

EWING, J. A. *An Engineer's Outlook*. Methuen, 1933.

EWING, A. W. *The Man of Room 40: Life of Sir Alfred Ewing*. Hutchinson, 1939.

GRANT, SIR A. *The Story of Edinburgh University during the first 300 years.* Longmans, 1884.

SHIPLEY, A. E. *'J', a memoir of John Willis Clark*. Smith, Elder, 1913.

SMALL, J. Engineering; a series of lectures delivered in the University of Glasgow, 1950–51, to commemorate the 5th centenary of the Foundation. *Fortuna Domus*, p. 335 (1950–1).

CHAPTER 5. BERTRAM HOPKINSON

EWING, J. A. Obituary Notice. *Proc. R. Soc.* A, **95** (1918–19), xxvi.

HOPKINSON, B. *Scientific Papers*. (Memoir by J. A. Ewing...) Cambridge University Press, 1921.

HOPKINSON, EVELYN. *The Story of a Mid Victorian Girl.* Cambridge University Press, 1928.

HOPKINSON, MARY and EWING, LADY. *John and Alice Hopkinson, 1824–1910.* Farmer and Sons, *c.* 1911.

HUDDLESTON, T. F. C. and HEYCOCK, C. T. Articles on National Security in *Nineteenth Century and After* (March, 1914).

PORTWAY, D. *Militant Don.* Robert Hale, 1964.

RICARDO, SIR H. R. Presidential Address. *Proc. Instn mech. Engrs*, **152** (1945), 143.

SHIPLEY, A. E. *'J', a memoir of John Willis Clark*. Smith, Elder, 1913.

CHAPTER 6. C. E. INGLIS

BAKER, J. F. Obituary Notice (of C. E. Inglis). Obituary Notices of Fellows, *Proc. R. Soc.* **8** (1953), 445.

BATCHELOR, C. K. (ed.). *The Scientific Papers of Sir Geoffrey Ingram Taylor.* Cambridge University Press, 1958.

INGLIS, C. E. Rehousing The Engineering Department. *Camb. Univ. Engng Soc. J.* **1**(1) (1921), 5.

INGLIS, C. E. Cambridge as a Place of Education. *Engineer, Lond.*, **152** (17 July 1931), 70.

INGLIS, C. E. Presidential Address. *J. Instn civ. Engrs*, **17** (1941–42), 1.

Obituary Notice, *King's College, Annual Report of the Council* (Nov. 1952).

PORTWAY, D. *Militant Don.* Robert Hale, 1964.

Royal Commission on Oxford and Cambridge Universities, 1922. Report. *Parliamentary Paper. 1922. Cmd 1588, X, 27.*

CHAPTER 7. J. F. BAKER

University and Departmental Records, Personal Reminiscences, etc.

INDEX

18-2